Periodic Table of the Elements

1 IA 1A	2 IIA 2A	3 IIIB 3B	4 IVB 4B	5 VB 5B	6 VIB 6B	7 VIIB 7B	8	9 VIII 8	10	11 IB 1B	12 IIB 2B	13 IIIA 3A	14 IVA 4A	15 VA 5A	16 VIA 6A	17 VIIA 7A	18 VIIIA 8A
1 **H** Hydrogen 1.008																	2 **He** Helium 4.003
3 **Li** Lithium 6.941	4 **Be** Beryllium 9.012											5 **B** Boron 10.811	6 **C** Carbon 12.011	7 **N** Nitrogen 14.007	8 **O** Oxygen 15.999	9 **F** Fluorine 18.998	10 **Ne** Neon 20.180
11 **Na** Sodium 22.990	12 **Mg** Magnesium 24.305											13 **Al** Aluminum 26.982	14 **Si** Silicon 28.086	15 **P** Phosphorus 30.974	16 **S** Sulfur 32.066	17 **Cl** Chlorine 35.453	18 **Ar** Argon 39.948
19 **K** Potassium 39.098	20 **Ca** Calcium 40.078	21 **Sc** Scandium 44.956	22 **Ti** Titanium 47.867	23 **V** Vanadium 50.942	24 **Cr** Chromium 51.996	25 **Mn** Manganese 54.938	26 **Fe** Iron 55.845	27 **Co** Cobalt 58.933	28 **Ni** Nickel 58.693	29 **Cu** Copper 63.546	30 **Zn** Zinc 65.38	31 **Ga** Gallium 69.723	32 **Ge** Germanium 72.631	33 **As** Arsenic 74.922	34 **Se** Selenium 78.971	35 **Br** Bromine 79.904	36 **Kr** Krypton 83.798
37 **Rb** Rubidium 85.468	38 **Sr** Strontium 87.62	39 **Y** Yttrium 88.906	40 **Zr** Zirconium 91.224	41 **Nb** Niobium 92.906	42 **Mo** Molybdenum 95.95	43 **Tc** Technetium 98.907	44 **Ru** Ruthenium 101.07	45 **Rh** Rhodium 102.906	46 **Pd** Palladium 106.42	47 **Ag** Silver 107.868	48 **Cd** Cadmium 112.414	49 **In** Indium 114.818	50 **Sn** Tin 118.711	51 **Sb** Antimony 121.760	52 **Te** Tellurium 127.6	53 **I** Iodine 126.904	54 **Xe** Xenon 131.294
55 **Cs** Cesium 132.905	56 **Ba** Barium 137.328	57-71	72 **Hf** Hafnium 178.49	73 **Ta** Tantalum 180.948	74 **W** Tungsten 183.84	75 **Re** Rhenium 186.207	76 **Os** Osmium 190.23	77 **Ir** Iridium 192.217	78 **Pt** Platinum 195.085	79 **Au** Gold 196.967	80 **Hg** Mercury 200.592	81 **Tl** Thallium 204.383	82 **Pb** Lead 207.2	83 **Bi** Bismuth 208.980	84 **Po** Polonium [208.982]	85 **At** Astatine 209.987	86 **Rn** Radon 222.018
87 **Fr** Francium 223.020	88 **Ra** Radium 226.025	89-103	104 **Rf** Rutherfordium [261]	105 **Db** Dubnium [262]	106 **Sg** Seaborgium [266]	107 **Bh** Bohrium [264]	108 **Hs** Hassium [269]	109 **Mt** Meitnerium [278]	110 **Ds** Darmstadtium [281]	111 **Rg** Roentgenium [280]	112 **Cn** Copernicium [285]	113 **Nh** Nihonium [286]	114 **Fl** Flerovium [289]	115 **Mc** Moscovium [289]	116 **Lv** Livermorium [293]	117 **Ts** Tennessine [294]	118 **Og** Oganesson [294]

Lanthanide Series	57 **La** Lanthanum 138.905	58 **Ce** Cerium 140.116	59 **Pr** Praseodymium 140.908	60 **Nd** Neodymium 144.243	61 **Pm** Promethium 144.913	62 **Sm** Samarium 150.36	63 **Eu** Europium 151.964	64 **Gd** Gadolinium 157.25	65 **Tb** Terbium 158.925	66 **Dy** Dysprosium 162.500	67 **Ho** Holmium 164.930	68 **Er** Erbium 167.259	69 **Tm** Thulium 168.934	70 **Yb** Ytterbium 173.055	71 **Lu** Lutetium 174.967
Actinide Series	89 **Ac** Actinium 227.028	90 **Th** Thorium 232.038	91 **Pa** Protactinium 231.036	92 **U** Uranium 238.029	93 **Np** Neptunium 237.048	94 **Pu** Plutonium 244.064	95 **Am** Americium 243.061	96 **Cm** Curium 247.070	97 **Bk** Berkelium 247.070	98 **Cf** Californium 251.080	99 **Es** Einsteinium [254]	100 **Fm** Fermium 257.095	101 **Md** Mendelevium 258.1	102 **No** Nobelium 259.101	103 **Lr** Lawrencium [262]

ISBN: 1733837795
ISBN-13: 978-1733837798

Hydrogen

Atomic Number: 1 **Atomic Symbol:** H
Category: Nonmetal **Standard State:** Gas

Physical & Atomic Properties

- Atomic Mass: 1.0080 u
- Electron Configuration: $1s^1$
- Oxidation States: +1, -1
- Electronegativity (Pauling Scale): 2.2
- Atomic Radius (van der Waals): 120 pm
- Ionization Energy: 13.598 eV
- Electron Affinity: 0.754 eV
- Melting Point: 13.81 K
- Boiling Point: 20.28 K
- Density: 0.00008988 g/cm^3

History & Characteristics

- Year Discovered: 1766
- Discovered by: Henry Cavendish
- Appearance: Colorless
- Natural Occurrence: Primordial
- Quick Fact: Hydrogen is the most abundant chemical substance in the universe.

Helium

Atomic Number: 2 **Atomic Symbol:** He
Category: Gas **Standard State:** Gas

Physical & Atomic Properties

- Atomic Mass: 4.00260 u
- Electron Configuration: $1s^2$
- Oxidation States: 0
- Atomic Radius (van der Waals): 140 pm
- Ionization Energy: 24.587 eV
- Melting Point: 0.95 K
- Boiling Point: 4.22 K
- Density: 0.0001785 g/cm^3

History & Characteristics

- Year Discovered: 1868
- Discovered by: Pierre Janssen
- Appearance: Colorless gas exhibiting a gray, cloudy glow when placed in an electric field.
- Natural Occurrence: Primordial
- Quick Fact: Helium has the lowest melting and boiling points of all the elements, and is the second most abundant chemical substance in the universe.

Lithium

Atomic Number: 3 **Atomic Symbol:** Li
Category: Alkali Metal **Standard State:** Solid

Physical & Atomic Properties

- Atomic Mass: 7.0 u
- Electron Configuration: [He]$2s^1$
- Oxidation States: +1
- Electronegativity (Pauling Scale): 0.98
- Atomic Radius (van der Waals): 182 pm
- Ionization Energy: 5.392 eV
- Electron Affinity: 0.618 eV
- Melting Point: 453.65 K
- Boiling Point: 1615 K
- Density: 0.534 g/cm^3

History & Characteristics

- Year Discovered: 1817
- Discovered by: Johan August Arfwedson
- Appearance: Silvery
- Natural Occurrence: Primordial
- Quick Fact: Lithium is so light it can float on water and is soft enough to be cut with a knife.

Beryllium

Atomic Number: 4 **Atomic Symbol:** Be
Category: Alkaline Earth Metal **Stand. State:** Solid

Physical & Atomic Properties

- Atomic Mass: 9.012183 u
- Electron Configuration: [He]$2s^2$
- Oxidation States: +2
- Electronegativity (Pauling Scale): 1.57
- Atomic Radius (van der Waals): 153 pm
- Ionization Energy: 9.323 eV
- Melting Point: 1560 K
- Boiling Point: 2744 K
- Density: 1.85 g/cm^3

History & Characteristics

- Year Discovered: 1798
- Discovered by: Louis Nicolas Vauquelin
- Appearance: Steel gray
- Natural Occurrence: Primordial
- Quick Fact: Beryllium is a rare element in the universe, usually occurring as a product of the spallation of larger atomic nuclei that have collided with cosmic rays.

Boron

Atomic Number: 5 **Atomic Symbol:** B
Category: Metalloid **Standard State:** Solid

Physical & Atomic Properties

- Atomic Mass: 10.81 u
- Electron Configuration: [He]$2s^2 2p^1$
- Oxidation States: +3
- Electronegativity (Pauling Scale): 2.04
- Atomic Radius (van der Waals): 192 pm
- Ionization Energy: 8.298 eV
- Electron Affinity: 0.277 eV
- Melting Point: 2348 K
- Boiling Point: 4273 K
- Density: 2.37 g/cm^3

History & Characteristics

- Year Discovered: 1808
- Discovered by: Joseph Louis Gay-Lussac and Louis-Jaques Thénard
- Appearance: Black
- Natural Occurrence: Primordial
- Quick Fact: Boron is found in small amounts in meteoroids, but chemically uncombined boron is not found naturally on Earth.

Carbon

Atomic Number: 6 **Atomic Symbol:** C
Category: Nonmetal **Standard State:** Solid

Physical & Atomic Properties

- Atomic Mass: 12.011 u
- Electron Configuration: $[He]2s^2 2p^2$
- Oxidation States: +4, +2, -4
- Electronegativity (Pauling Scale): 2.55
- Atomic Radius (van der Waals): 170 pm
- Ionization Energy: 11.260 eV
- Electron Affinity: 1.263 eV
- Melting Point: 3823 K
- Boiling Point: 4098 K
- Density: 2.2670 g/cm^3

History & Characteristics

- Year Discovered: Ancient
- Discovered by: Sumerians (presumed)
- Appearance: Graphite is opaque and black; Diamond is highly transparent.
- Natural Occurrence: Primordial
- Quick Fact: Carbon is known to form nearly 10 million compounds, and life on Earth is generally referred to as "carbon-based life."

Nitrogen

Atomic Number: 7 **Atomic Symbol:** N
Category: Nonmetal **Standard State:** Gas

Physical & Atomic Properties

- Atomic Mass: 14.007 u
- Electron Configuration: [He]$2s^2 2p^3$
- Oxidation States: +5, +4, +3, +2, +1, -1, -2, -3
- Electronegativity (Pauling Scale): 3.04
- Atomic Radius (van der Waals): 155 pm
- Ionization Energy: 14.534 eV
- Melting Point: 63.15 K
- Boiling Point: 77.36 K
- Density: 0.0012506 g/cm^3

History & Characteristics

- Year Discovered: 1772
- Discovered by: Daniel Rutherford
- Appearance: Colorless
- Natural Occurrence: Primordial
- Quick Fact: Approximately 78% of air in the Earth's atmosphere is made up of nitrogen.

Oxygen

Atomic Number: 8 **Atomic Symbol:** O
Category: Nonmetal **Standard State:** Gas

Physical & Atomic Properties

- Atomic Mass: 15.999 u
- Electron Configuration: [He]$2s^2 2p^4$
- Oxidation States: -2
- Electronegativity (Pauling Scale): 3.44
- Atomic Radius (van der Waals): 152 pm
- Ionization Energy: 13.618 eV
- Electron Affinity: 1.461 eV
- Melting Point: 54.36 K
- Boiling Point: 90.2 K
- Density: 0.001429 g/cm^3

History & Characteristics

- Year Discovered: 1772
- Discovered by: Carl Wilhelm Scheele
- Appearance: Colorless
- Natural Occurrence: Primordial
- Quick Fact: Oxygen is continuously replenished in the Earth's atmosphere through the process of photosynthesis.

Fluorine

Atomic Number: 9 **Atomic Symbol:** F
Category: Halogen **Standard State:** Gas

Physical & Atomic Properties

- Atomic Mass: 18.99840316 u
- Electron Configuration: [He]$2s^2 2p^5$
- Oxidation States: -1
- Electronegativity (Pauling Scale): 3.98
- Atomic Radius (van der Waals): 135 pm
- Ionization Energy: 17.423 eV
- Electron Affinity: 3.339 eV
- Melting Point: 53.53 K
- Boiling Point: 85.03 K
- Density: 0.001696 g/cm^3

History & Characteristics

- Year Discovered: 1886
- Discovered by: Henri Moissan
- Appearance: Pale yellow
- Natural Occurrence: Primordial
- Quick Fact: Fluorine is the most electronegative element, and the bond formed between fluorine and carbon is the strongest bond in organic chemistry.

Neon

Atomic Number: 10 **Atomic Symbol:** Ne
Category: Noble Gas **Standard State:** Gas

Physical & Atomic Properties

- Atomic Mass: 20.180 u
- Electron Configuration: [He]$2s^2 2p^6$
- Oxidation States: 0
- Atomic Radius (van der Waals): 154 pm
- Ionization Energy: 21.565 eV
- Melting Point: 24.56 K
- Boiling Point: 27.07 K
- Density: 0.0008999 g/cm^3

History & Characteristics

- Year Discovered: 1898
- Discovered by: William Ramsay and Morris W. Travers
- Appearance: Colorless
- Natural Occurrence: Primordial
- Quick Fact: Neon is an inert monatomic gas, meaning its atoms don't bond like oxygen and nitrogen to form compounds. This makes neon gas "lighter than air."

Sodium

Atomic Number: 11 **Atomic Symbol:** Na
Category: Alkali Metal **Standard State:** Solid

Physical & Atomic Properties

- Atomic Mass: 22.9897693 u
- Electron Configuration: [Ne]$3s^1$
- Oxidation States: +1
- Electronegativity (Pauling Scale): 0.93
- Atomic Radius (van der Waals): 227 pm
- Ionization Energy: 5.139 eV
- Electron Affinity: 0.548 eV
- Melting Point: 370.95 K
- Boiling Point: 1156 K
- Density: 0.97 g/cm^3

History & Characteristics

- Year Discovered: 1807
- Discovered by: Humphry Davy
- Appearance: Silvery-white
- Natural Occurrence: Primordial
- Quick Fact: Sodium is an essential element for all animals, and helps the human body maintain proper fluid balance in its cells.

Magnesium

Atomic Number: 12 **Atomic Symbol:** Mg
Category: Alkaline Earth Metal **Stand. State:** Solid

Physical & Atomic Properties

- Atomic Mass: 24.305 u
- Electron Configuration: [Ne]$3s^2$
- Oxidation States: +2
- Electronegativity (Pauling Scale): 1.31
- Atomic Radius (van der Waals): 173 pm
- Ionization Energy: 7.646 eV
- Melting Point: 923 K
- Boiling Point: 1363 K
- Density: 1.74 g/cm^3

History & Characteristics

- Year Discovered: 1755
- Discovered by: Joseph Black
- Appearance: Silvery-white
- Natural Occurrence: Primordial
- Quick Fact: Magnesium is commonly used in flares and fireworks because of the bright white light it produces when it burns.

Aluminum

Atomic Number: 13 **Atomic Symbol:** Al
Cat: Post-Transition Metal **Stand. State:** Solid

Physical & Atomic Properties

- Atomic Mass: 26.981538 u
- Electron Configuration: [Ne]$3s^2 3p^1$
- Oxidation States: +3
- Electronegativity (Pauling Scale): 1.61
- Atomic Radius (van der Waals): 184 pm
- Ionization Energy: 5.986 eV
- Electron Affinity: 0.441 eV
- Melting Point: 933.437 K
- Boiling Point: 2792 K
- Density: 2.70 g/cm^3

History & Characteristics

- Year Discovered: 1825
- Discovered by: Hans Christian Ørsted
- Appearance: Silvery
- Natural Occurrence: Primordial
- Quick Fact: Aluminum is remarkable for its low density and ability to resist corrosion, making it vital to the aerospace, building, and transportation industries.

Silicon

Atomic Number: 14
Atomic Symbol: Si
Category: Metalloid
Standard State: Solid

Physical & Atomic Properties

- Atomic Mass: 28.085 u
- Electron Configuration: [Ne]$3s^2 3p^2$
- Oxidation States: +4, +2, -4
- Electronegativity (Pauling Scale): 1.9
- Atomic Radius (van der Waals): 210 pm
- Ionization Energy: 8.152 eV
- Electron Affinity: 1.385 eV
- Melting Point: 1687 K
- Boiling Point: 3538 K
- Density: 2.3296 g/cm^3

History & Characteristics

- Year Discovered: 1824
- Discovered by: Jöns Jacob Berzelius
- Appearance: Silvery
- Natural Occurrence: Primordial
- Quick Fact: Silicon acts as a semiconductor, making it an essential part of electronic devices from radios to cell phones.

Phosphorus

Atomic Number: 15 **Atomic Symbol:** P
Category: Nonmetal **Standard State:** Solid

Physical & Atomic Properties

- Atomic Mass: 30.97376200 u
- Electron Configuration: [Ne]$3s^2 3p^3$
- Oxidation States: +5, +3, -3
- Electronegativity (Pauling Scale): 2.19
- Atomic Radius (van der Waals): 180 pm
- Ionization Energy: 10.487 eV
- Electron Affinity: 0.746 eV
- Melting Point: 317.3 K
- Boiling Point: 553.65 K
- Density: 1.82 g/cm^3

History & Characteristics

- Year Discovered: 1669
- Discovered by: Hennig Brand
- Appearance: Varies; generally colorless or waxy white, yellow, red, violet, or black.
- Natural Occurrence: Primordial
- Quick Fact: Phosphorus is essential to life, as phosphates are a major part of each molecule of DNA and RNA.

Sulfur

Atomic Number: 16 **Atomic Symbol:** S
Category: Nonmetal **Standard State:** Solid

Physical & Atomic Properties

- Atomic Mass: 32.07 u
- Electron Configuration: [Ne]$3s^2 3p^4$
- Oxidation States: +6, +4, -2
- Electronegativity (Pauling Scale): 2.58
- Atomic Radius (van der Waals): 180 pm
- Ionization Energy: 10.360 eV
- Electron Affinity: 2.077 eV
- Melting Point: 388.36 K
- Boiling Point: 717.75 K
- Density: 2.067 g/cm^3

History & Characteristics

- Year Discovered: Ancient
- Discovered by: Chinese (presumed)
- Appearance: Yellow
- Natural Occurrence: Primordial
- Quick Fact: Many sulfur compounds are odoriferous, and the smells of odorized natural gas and skunk scent are due to organosulfur compounds.

Chlorine

Atomic Number: 17 **Atomic Symbol:** Cl
Category: Halogen **Standard State:** Gas

<u>**Physical & Atomic Properties**</u>

- Atomic Mass: 35.45 u
- Electron Configuration: [Ne]$3s^2 3p^5$
- Oxidation States: +7, +5, +1, -1
- Electronegativity (Pauling Scale): 3.16
- Atomic Radius (van der Waals): 175 pm
- Ionization Energy: 12.968 eV
- Electron Affinity: 3.617 eV
- Melting Point: 171.65 K
- Boiling Point: 239.11 K
- Density: 0.003214 g/cm^3

<u>**History & Characteristics**</u>

- Year Discovered: 1774
- Discovered by: Carl Wilhelm Scheele
- Appearance: Greenish-yellow
- Natural Occurrence: Primordial
- Quick Fact: There is so much chlorine in the Earth's oceans that it would weigh five times more than our present atmosphere if it were released as a gas.

Argon

Atomic Number: 18 **Atomic Symbol:** Ar
Category: Noble Gas **Standard State:** Gas

Physical & Atomic Properties

- Atomic Mass: 39.9 u
- Electron Configuration: [Ne]$3s^2 3p^6$
- Oxidation States: 0
- Atomic Radius (van der Waals): 188 pm
- Ionization Energy: 15.760 eV
- Melting Point: 83.8 K
- Boiling Point: 87.3 K
- Density: 0.0017837 g/cm^3

History & Characteristics

- Year Discovered: 1894
- Discovered by: Lord Rayleigh and William Ramsay
- Appearance: Colorless
- Natural Occurrence: Primordial
- Quick Fact: Argon is the third most abundant gas in the Earth's atmosphere, and the most abundant noble gas in the Earth's crust.

Potassium

Atomic Number: 19 **Atomic Symbol:** K
Category: Metal **Standard State:** Solid

Physical & Atomic Properties

- Atomic Mass: 39.098 u
- Electron Configuration: $[Ar]4s^1$
- Oxidation States: +1
- Electronegativity (Pauling Scale): 0.82
- Atomic Radius (van der Waals): 275 pm
- Ionization Energy: 4.341 eV
- Electron Affinity: 0.501 eV
- Melting Point: 336.53 K
- Boiling Point: 1032 K
- Density: 0.89 g/cm^3

History & Characteristics

- Year Discovered: 1807
- Discovered by: Humphry Davy
- Appearance: Silvery-white
- Natural Occurrence: Primordial
- Quick Fact: Potassium is the second lightest metal after lithium, and is soft enough to be cut with a knife with little force.

Calcium

Atomic Number: 20 **Atomic Symbol:** Ca
Category: Alkaline Earth Metal **Stand. State:** Solid

Physical & Atomic Properties

- Atomic Mass: 40.08 u
- Electron Configuration: [Ar]$4s^2$
- Oxidation States: +2
- Electronegativity (Pauling Scale): 1
- Atomic Radius (van der Waals): 231 pm
- Ionization Energy: 6.113 eV
- Melting Point: 1115 K
- Boiling Point: 1757 K
- Density: 1.54 g/cm^3

History & Characteristics

- Year Discovered: 1808
- Discovered by: Humphry Davy
- Appearance: Silvery-gray
- Natural Occurrence: Primordial
- Quick Fact: The average adult body contains about 2 lb. of calcium, 99% of which is in the bones and teeth, making it the most abundant of the metallic elements in the human body.

Scandium

Atomic Number: 21 **Atomic Symbol:** Sc
Category: Transition Metal **Stand. State:** Solid

Physical & Atomic Properties

- Atomic Mass: 44.95591 u
- Electron Configuration: [Ar]$4s^2 3d^1$
- Oxidation States: +3
- Electronegativity (Pauling Scale): 1.36
- Atomic Radius (van der Waals): 211 pm
- Ionization Energy: 6.561 eV
- Electron Affinity: 0.188 eV
- Melting Point: 1814 K
- Boiling Point: 3109 K
- Density: 2.99 g/cm^3

History & Characteristics

- Year Discovered: 1879
- Discovered by: Lars Fredrik Nilson
- Appearance: Silvery-white
- Natural Occurrence: Primordial
- Quick Fact: Scandium is considered a rare-earth element and is the first of the transition metals. It's usually obtained as a byproduct during uranium refinement.

Titanium

Atomic Number: 22 **Atomic Symbol:** Ti
Category: Transition Metal **Stand. State:** Solid

Physical & Atomic Properties

- Atomic Mass: 47.87 u
- Electron Configuration: $[Ar]4s^23d^2$
- Oxidation States: +4, +3, +2
- Electronegativity (Pauling Scale): 1.54
- Atomic Radius (van der Waals): 187 pm
- Ionization Energy: 6.828 eV
- Electron Affinity: 0.079 eV
- Melting Point: 1941 K
- Boiling Point: 3560 K
- Density: 4.5 g/cm^3

History & Characteristics

- Year Discovered: 1791
- Discovered by: William Gregor
- Appearance: Silvery-white
- Natural Occurrence: Primordial
- Quick Fact: Titanium is corrosion resistant (even from water and chlorine) and has the highest strength-to-density ratio of any metallic element.

Vanadium

Atomic Number: 23 **Atomic Symbol:** V
Category: Transition Metal **Stand. State:** Solid

Physical & Atomic Properties

- Atomic Mass: 50.941 u
- Electron Configuration: [Ar]$4s^2 3d^3$
- Oxidation States: +5, +4, +3, +2
- Electronegativity (Pauling Scale): 1.63
- Atomic Radius (van der Waals): 179 pm
- Ionization Energy: 6.746 eV
- Electron Affinity: 0.525 eV
- Melting Point: 2183 K
- Boiling Point: 3680 K
- Density: 6.0 g/cm^3

History & Characteristics

- Year Discovered: 1801
- Discovered by: Andres Manuel del Rio
- Appearance: Silvery
- Natural Occurrence: Primordial
- Quick Fact: Vanadium is corrosion-resistant and malleable, making it useful in the manufacturing industry. One of its first uses was in the steel chassis of the 1908 Model T Ford.

Chromium

Atomic Number: 24 **Atomic Symbol:** Cr
Category: Transition Metal **Stand. State:** Solid

Physical & Atomic Properties

- Atomic Mass: 51.996 u
- Electron Configuration: $[Ar]3d^5 4s^1$
- Oxidation States: +6, +3, +2
- Electronegativity (Pauling Scale): 1.66
- Atomic Radius (van der Waals): 189 pm
- Ionization Energy: 6.767 eV
- Electron Affinity: 0.666 eV
- Melting Point: 2180 K
- Boiling Point: 2944 K
- Density: 7.15 g/cm^3

History & Characteristics

- Year Discovered: 1797
- Discovered by: Louis Nicolas Vauquelin
- Appearance: Silvery-gray
- Natural Occurrence: Primordial
- Quick Fact: Chromium is the main additive in stainless steel, to which it adds anti-corrosive properties.

Manganese

Atomic Number: 25 **Atomic Symbol:** Mn
Category: Transition Metal **Stand. State:** Solid

Physical & Atomic Properties

- Atomic Mass: 54.93804 u
- Electron Configuration: $[Ar]4s^2 3d^5$
- Oxidation States: +7, +4, +3, +2
- Electronegativity (Pauling Scale): 1.55
- Atomic Radius (van der Waals): 197 pm
- Ionization Energy: 7.434 eV
- Melting Point: 1519 K
- Boiling Point: 2334 K
- Density: 7.3 g/cm^3

History & Characteristics

- Year Discovered: 1774
- Discovered by: Carl Wilhelm Scheele
- Appearance: Gray-white
- Natural Occurrence: Primordial
- Quick Fact: The human body contains about 12 mg of manganese, and it's present as a coenzyme in biological processes including macronutrient metabolism, bone formation, and wound healing.

Iron

Atomic Number: 26 **Atomic Symbol:** Fe
Category: Transition Metal **Stand. State:** Solid

Physical & Atomic Properties

- Atomic Mass: 55.84 u
- Electron Configuration: [Ar]$4s^2 3d^6$
- Oxidation States: +3, +2
- Electronegativity (Pauling Scale): 1.83
- Atomic Radius (van der Waals): 194 pm
- Ionization Energy: 7.902 eV
- Electron Affinity: 0.163 eV
- Melting Point: 1811 K
- Boiling Point: 3134 K
- Density: 7.874 g/cm^3

History & Characteristics

- Year Discovered: Ancient
- Discovered by: Mesopotamians (presumed)
- Appearance: Silvery-gray
- Natural Occurrence: Primordial
- Quick Fact: By mass, iron is the most common element on Earth, forming much of Earth's outer and inner core.

Cobalt

Atomic Number: 27 **Atomic Symbol:** Co
Category: Transition Metal **Stand. State:** Solid

Physical & Atomic Properties

- Atomic Mass: 58.93319 u
- Electron Configuration: [Ar]$4s^2 3d^7$
- Oxidation States: +3, +2
- Electronegativity (Pauling Scale): 1.88
- Atomic Radius (van der Waals): 192 pm
- Ionization Energy: 7.881 eV
- Electron Affinity: 0.661 eV
- Melting Point: 1768 K
- Boiling Point: 3200 K
- Density: 8.86 g/cm^3

History & Characteristics

- Year Discovered: 1735
- Discovered by: Georg Brandt
- Appearance: Bluish-white
- Natural Occurrence: Primordial
- Quick Fact: Cobalt is primarily used in lithium-ion batteries, and in the manufacturing of magnetic, wear-resistant, high-strength alloys.

Nickel

Atomic Number: 28
Atomic Symbol: Ni
Category: Transition Metal
Stand. State: Solid

Physical & Atomic Properties

- Atomic Mass: 58.693 u
- Electron Configuration: [Ar]$4s^2 3d^8$
- Oxidation States: +3, +2
- Electronegativity (Pauling Scale): 1.91
- Atomic Radius (van der Waals): 163 pm
- Ionization Energy: 7.640 eV
- Electron Affinity: 1.156 eV
- Melting Point: 1728 K
- Boiling Point: 3186 K
- Density: 8.912 g/cm^3

History & Characteristics

- Year Discovered: 1751
- Discovered by: Axel Fredrik Cronstedt
- Appearance: Silvery-white
- Natural Occurrence: Primordial
- Quick Fact: The U.S. five cent coin, the "nickel," is composed of 75% copper and 25% nickel.

Copper

Atomic Number: 29 **Atomic Symbol:** Cu
Category: Transition Metal **Stand. State:** Solid

Physical & Atomic Properties

- Atomic Mass: 63.55 u
- Electron Configuration: [Ar]$4s^1 3d^{10}$
- Oxidation States: +2, +1
- Electronegativity (Pauling Scale): 1.9
- Atomic Radius (van der Waals): 140 pm
- Ionization Energy: 7.726 eV
- Electron Affinity: 1.228 eV
- Melting Point: 1357.77 K
- Boiling Point: 2835 K
- Density: 8.933 g/cm^3

History & Characteristics

- Year Discovered: Ancient
- Discovered by: Mesopotamians (presumed)
- Appearance: Orange-red
- Natural Occurrence: Primordial
- Quick Fact: The largest single piece of native copper ever found weighed over 520 tons, but most copper ore that is mined only contains about 1% of the metal.

Zinc

Atomic Number: 30 **Atomic Symbol:** Zn
Category: Transition Metal **Stand. State:** Solid

Physical & Atomic Properties

- Atomic Mass: 65.4 u
- Electron Configuration: [Ar]$4s^2 3d^{10}$
- Oxidation States: +2
- Electronegativity (Pauling Scale): 1.65
- Atomic Radius (van der Waals): 139 pm
- Ionization Energy: 9.394 eV
- Melting Point: 692.68 K
- Boiling Point: 1180 K
- Density: 7.134 g/cm^3

History & Characteristics

- Year Discovered: Ancient
- Discovered by: Ancient India (presumed)
- Appearance: Bluish-silver
- Natural Occurrence: Primordial
- Quick Fact: The human body contains between 2 to 4 grams of zinc, and it is considered an essential mineral. Common foods that contain zinc are sesame seeds, wheat, beans, nuts, and mustard.

Gallium

Atomic Number: 31 **Atomic Symbol:** Ga
Cat: Post-Transition Metal **Stand. State:** Solid

Physical & Atomic Properties

- Atomic Mass: 69.72 u
- Electron Configuration: [Ar]$4s^2 3d^{10} 4p^1$
- Oxidation States: +3
- Electronegativity (Pauling Scale): 1.81
- Atomic Radius (van der Waals): 187 pm
- Ionization Energy: 5.999 eV
- Electron Affinity: 0.3 eV
- Melting Point: 302.91 K
- Boiling Point: 2477 K
- Density: 5.91 g/cm^3

History & Characteristics

- Year Discovered: 1875
- Discovered by: Paul-Émile Lecoq de Boisbaudran
- Appearance: Silvery-blue
- Natural Occurrence: Primordial
- Quick Fact: Gallium has no known natural role in biology. It's primarily used as part of a chemical compound in electronics.

Germanium

Atomic Number: 32 **Atomic Symbol:** Ge
Category: Metalloid **Standard State:** Solid

Physical & Atomic Properties

- Atomic Mass: 72.63 u
- Electron Configuration: $[Ar]4s^2 3d^{10} 4p^2$
- Oxidation States: +4, +2
- Electronegativity (Pauling Scale): 2.01
- Atomic Radius (van der Waals): 211 pm
- Ionization Energy: 7.900 eV
- Electron Affinity: 1.35 eV
- Melting Point: 1211.4 K
- Boiling Point: 3106 K
- Density: 5.323 g/cm^3

History & Characteristics

- Year Discovered: 1886
- Discovered by: Clemens Winkler
- Appearance: Gray-white
- Natural Occurrence: Primordial
- Quick Fact: Germanium is used as a semiconductor in transistors and electronic devices. The first decade of semiconductor electronics was based entirely on germanium.

Arsenic

Atomic Number: 33 **Atomic Symbol:** As
Category: Metalloid **Standard State:** Solid

Physical & Atomic Properties

- Atomic Mass: 74.92159 u
- Electron Configuration: $[Ar]4s^2 3d^{10} 4p^3$
- Oxidation States: +5, +3, -3
- Electronegativity (Pauling Scale): 2.18
- Atomic Radius (van der Waals): 185 pm
- Ionization Energy: 9.815 eV
- Electron Affinity: 0.81 eV
- Melting Point: 1090 K
- Boiling Point: 887 K
- Density: 5.776 g/cm^3

History & Characteristics

- Year Discovered: 1250
- Discovered by: Albertus Magnus
- Appearance: Gray
- Natural Occurrence: Primordial
- Quick Fact: Arsenic is classified as a Group-A carcinogen, and the U.S. Environmental Protection Agency states that all forms of arsenic are a serious risk to human health.

Selenium

Atomic Number: 34 **Atomic Symbol:** Se
Category: Nonmetal **Standard State:** Solid

Physical & Atomic Properties

- Atomic Mass: 78.97 u
- Electron Configuration: [Ar]$4s^2 3d^{10} 4p^4$
- Oxidation States: +6, +4, -2
- Electronegativity (Pauling Scale): 2.55
- Atomic Radius (van der Waals): 190 pm
- Ionization Energy: 9.752 eV
- Electron Affinity: 2.021 eV
- Melting Point: 493.65 K
- Boiling Point: 958 K
- Density: 4.809 g/cm^3

History & Characteristics

- Year Discovered: 1817
- Discovered by: Jöns Jacob Berzelius
- Appearance: Gray or red (crystalline), black or red (amorphous)
- Natural Occurrence: Primordial
- Quick Fact: The chief uses of selenium are in glassmaking and pigments, but it's also a semiconductor commonly used in photocells.

Bromine

Atomic Number: 35 **Atomic Symbol:** Br
Category: Halogen **Standard State:** Liquid

Physical & Atomic Properties

- Atomic Mass: 79.90 u
- Electron Configuration: [Ar]$4s^2 3d^{10} 4p^5$
- Oxidation States: +5, +1, -1
- Electronegativity (Pauling Scale): 2.96
- Atomic Radius (van der Waals): 183 pm
- Ionization Energy: 11.814 eV
- Electron Affinity: 3.365 eV
- Melting Point: 265.95 K
- Boiling Point: 331.95 K
- Density: 3.11 g/cm^3

History & Characteristics

- Year Discovered: 1826
- Discovered by: Antoine-Jérôme Balard
- Appearance: Red-brown
- Natural Occurrence: Primordial
- Quick Fact: Bromine is used in many fire retardant compounds. When brominated compounds burn, hydrobromic acid is produced which acts as a flame retardant.

Krypton

Atomic Number: 36 **Atomic Symbol:** Kr
Category: Gas **Standard State:** Gas

Physical & Atomic Properties

- Atomic Mass: 83.80 u
- Electron Configuration: $[Ar]4s^2 3d^{10} 4p^6$
- Oxidation States: 0
- Electronegativity (Pauling Scale): 3
- Atomic Radius (van der Waals): 202 pm
- Ionization Energy: 14.000 eV
- Melting Point: 115.79 K
- Boiling Point: 119.93 K
- Density: 0.003733 g/cm^3

History & Characteristics

- Year Discovered: 1898
- Discovered by: William Ramsay and Morris W. Travers
- Appearance: Colorless
- Natural Occurrence: Primordial
- Quick Fact: Krypton is injected into incandescent lightbulbs to extend the life of the tungsten filament that makes those bulbs glow.

Rubidium

Atomic Number: 37
Atomic Symbol: Rb
Category: Alkali Metal
Standard State: Solid

Physical & Atomic Properties

- Atomic Mass: 85.468 u
- Electron Configuration: [Kr]$5s^1$
- Oxidation States: +1
- Electronegativity (Pauling Scale): 0.82
- Atomic Radius (van der Waals): 303 pm
- Ionization Energy: 4.177 eV
- Electron Affinity: 0.468 eV
- Melting Point: 312.46 K
- Boiling Point: 961 K
- Density: 1.53 g/cm^3

History & Characteristics

- Year Discovered: 1861
- Discovered by: Robert Bunsen and Gustav Kirchhoff
- Appearance: Silvery-white
- Natural Occurrence: Primordial
- Quick Fact: Rubidium ignites when exposed to air and reacts violently with water, releasing hydrogen, which bursts into flames.

Strontium

Atomic Number: 38 **Atomic Symbol:** Sr
Category: Alkaline Earth Metal **Stand. State:** Solid

Physical & Atomic Properties

- Atomic Mass: 87.6 u
- Electron Configuration: [Kr]$5s^2$
- Oxidation States: +2
- Electronegativity (Pauling Scale): 0.95
- Atomic Radius (van der Waals): 249 pm
- Ionization Energy: 5.695 eV
- Melting Point: 1050 K
- Boiling Point: 1655 K
- Density: 2.64 g/cm^3

History & Characteristics

- Year Discovered: 1790
- Discovered by: Adair Crawford
- Appearance: Silvery
- Natural Occurrence: Primordial
- Quick Fact: While natural strontium is stable, the synthetic form, strontium-90, is radio-active and is one of the most dangerous components of nuclear fallout.

Yttrium

Atomic Number: 39 **Atomic Symbol:** Y
Category: Transition Metal **Stand. State:** Solid

Physical & Atomic Properties

- Atomic Mass: 88.9058 u
- Electron Configuration: [Kr]$5s^2 4d^1$
- Oxidation States: +3
- Electronegativity (Pauling Scale): 1.22
- Atomic Radius (van der Waals): 219 pm
- Ionization Energy: 6.217 eV
- Electron Affinity: 0.307 eV
- Melting Point: 1795 K
- Boiling Point: 3618 K
- Density: 4.47 g/cm^3

History & Characteristics

- Year Discovered: 1794
- Discovered by: Johan Gadolin
- Appearance: Silvery-white
- Natural Occurrence: Primordial
- Quick Fact: Yttrium is used to increase the strength of aluminum and magnesium alloys, and yttrium-90, a radioactive isotope, is used to treat various cancers.

Zirconium

Atomic Number: 40
Atomic Symbol: Zr
Category: Transition Metal
Stand. State: Solid

Physical & Atomic Properties

- Atomic Mass: 91.22 u
- Electron Configuration: [Kr]$5s^2 4d^2$
- Oxidation States: +4
- Electronegativity (Pauling Scale): 1.33
- Atomic Radius (van der Waals): 186 pm
- Ionization Energy: 6.634 eV
- Electron Affinity: 0.426 eV
- Melting Point: 2128 K
- Boiling Point: 4682 K
- Density: 6.52 g/cm^3

History & Characteristics

- Year Discovered: 1789
- Discovered by: Martin Heinrich Klaproth
- Appearance: Grayish-white
- Natural Occurrence: Primordial
- Quick Fact: Because zirconium is resistant to corrosion it is commonly used as an alloying agent in materials like surgical appliances, light filaments, and watch cases.

Niobium

Atomic Number: 41 **Atomic Symbol:** Nb
Category: Transition Metal **Stand. State:** Solid

Physical & Atomic Properties

- Atomic Mass: 92.9064 u
- Electron Configuration: [Kr]$5s^1 4d^4$
- Oxidation States: +5, +3
- Electronegativity (Pauling Scale): 1.6
- Atomic Radius (van der Waals): 207 pm
- Ionization Energy: 6.759 eV
- Electron Affinity: 0.893 eV
- Melting Point: 2750 K
- Boiling Point: 5017 K
- Density: 8.57 g/cm^3

History & Characteristics

- Year Discovered: 1801
- Discovered by: Charles Hatchett
- Appearance: Silvery-white
- Natural Occurrence: Primordial
- Quick Fact: Nearly 80% of niobium is used in the steel industry to create high-strength, low-alloy steels.

Molybdenum

Atomic Number: 42 **Atomic Symbol:** Mo
Category: Transition Metal **Stand. State:** Solid

Physical & Atomic Properties

- Atomic Mass: 96.0 u
- Electron Configuration: [Kr]$5s^1 4d^5$
- Oxidation States: +6
- Electronegativity (Pauling Scale): 2.16
- Atomic Radius (van der Waals): 209 pm
- Ionization Energy: 7.092 eV
- Electron Affinity: 0.746 eV
- Melting Point: 2896 K
- Boiling Point: 4912 K
- Density: 10.2 g/cm^3

History & Characteristics

- Year Discovered: 1778
- Discovered by: Carl Wilhelm Scheele
- Appearance: Silvery-white
- Natural Occurrence: Primordial
- Quick Fact: Molybdenum is used in the production of alloys, where it is added to increase hardness, strength, electrical conductivity and resistance to corrosion.

Technetium

Atomic Number: 43 **Atomic Symbol:** Tc
Category: Transition Metal **Stand. State:** Solid

Physical & Atomic Properties

- Atomic Mass: 97.90721 u
- Electron Configuration: [Kr]$5s^2 4d^5$
- Oxidation States: +7, +6, +4
- Electronegativity (Pauling Scale): 1.9
- Atomic Radius (van der Waals): 209 pm
- Ionization Energy: 7.28 eV
- Electron Affinity: 0.55 eV
- Melting Point: 2430 K
- Boiling Point: 4538 K
- Density: 11 g/cm^3

History & Characteristics

- Year Discovered: 1937
- Discovered by: Carlo Perrier and Emilio Segrè
- Appearance: Silvery-gray
- Natural Occurrence: From decay
- Quick Fact: Technetium is primarily obtained artificially, and every form of technetium is radioactive.

Ruthenium

Atomic Number: 44 **Atomic Symbol:** Ru
Category: Transition Metal **Stand. State:** Solid

Physical & Atomic Properties

- Atomic Mass: 101.1 u
- Electron Configuration: [Kr]$5s^1 4d^7$
- Oxidation States: +3
- Electronegativity (Pauling Scale): 2.2
- Atomic Radius (van der Waals): 207 pm
- Ionization Energy: 7.361 eV
- Electron Affinity: 1.05 eV
- Melting Point: 2607 K
- Boiling Point: 4423 K
- Density: 12.1 g/cm^3

History & Characteristics

- Year Discovered: 1844
- Discovered by: Karl Karlovich Klaus
- Appearance: Silvery-white
- Natural Occurrence: Primordial
- Quick Fact: Most ruthenium produced is used in wear-resistant electrical contacts and thick-film resistors.

Rhodium

Atomic Number: 45
Atomic Symbol: Rh
Category: Transition Metal
Stand. State: Solid

Physical & Atomic Properties

- Atomic Mass: 102.9055 u
- Electron Configuration: [Kr]$5s^1 4d^8$
- Oxidation States: +3
- Electronegativity (Pauling Scale): 2.28
- Atomic Radius (van der Waals): 195 pm
- Ionization Energy: 7.459 eV
- Electron Affinity: 1.137 eV
- Melting Point: 2237 K
- Boiling Point: 3968 K
- Density: 12.4 g/cm^3

History & Characteristics

- Year Discovered: 1803
- Discovered by: William Hyde Wollaston
- Appearance: Silvery-white
- Natural Occurrence: Primordial
- Quick Fact: Rhodium is considered the rarest and most valuable precious metal in the world—well above gold or silver.

Palladium

Atomic Number: 46 **Atomic Symbol:** Pd

Category: Transition Metal **Stand. State:** Solid

Physical & Atomic Properties

- Atomic Mass: 106.4 u
- Electron Configuration: [Kr]$4d^{10}$
- Oxidation States: +3, +2
- Electronegativity (Pauling Scale): 2.2
- Atomic Radius (van der Waals): 202 pm
- Ionization Energy: 8.337 eV
- Electron Affinity: 0.557 eV
- Melting Point: 1828.05 K
- Boiling Point: 3236 K
- Density: 12.0 g/cm^3

History & Characteristics

- Year Discovered: 1803
- Discovered by: William Hyde Wollaston
- Appearance: Silvery-white
- Natural Occurrence: Primordial
- Quick Fact: Palladium is extremely ductile and its chief use is in automobile catalytic converters.

Silver

Atomic Number: 47 **Atomic Symbol:** Ag
Category: Transition Metal **Stand. State:** Solid

Physical & Atomic Properties

- Atomic Mass: 107.868 u
- Electron Configuration: $[Kr]5s^1 4d^{10}$
- Oxidation States: +1
- Electronegativity (Pauling Scale): 1.93
- Atomic Radius (van der Waals): 172 pm
- Ionization Energy: 7.576 eV
- Electron Affinity: 1.302 eV
- Melting Point: 1234.93 K
- Boiling Point: 2435 K
- Density: 10.501 g/cm^3

History & Characteristics

- Year Discovered: Ancient
- Discovered by: Various ancient civilizations
- Appearance: Silver
- Natural Occurrence: Primordial
- Quick Fact: Silver exhibits the highest electrical conductivity, thermal conductivity, and reflectivity of any metal.

Cadmium

Atomic Number: 48 **Atomic Symbol:** Cd
Category: Transition Metal **Stand. State:** Solid

Physical & Atomic Properties

- Atomic Mass: 112.41 u
- Electron Configuration: [Kr]$5s^2 4d^{10}$
- Oxidation States: +2
- Electronegativity (Pauling Scale): 1.69
- Atomic Radius (van der Waals): 158 pm
- Ionization Energy: 8.994 eV
- Melting Point: 594.22 K
- Boiling Point: 1040 K
- Density: 8.69 g/cm^3

History & Characteristics

- Year Discovered: 1817
- Discovered by: Friedrich Stromeyer
- Appearance: Bluish-white
- Natural Occurrence: Primordial
- Quick Fact: Cadmium was used for many years as a corrosion-resistant plating on steel, and its compounds are used as red, orange, and yellow pigments to color glass and stabilize plastic.

Indium

Atomic Number: 49 **Atomic Symbol:** In
Cat: Post-Transition Metal **Stand. State:** Solid

Physical & Atomic Properties

- Atomic Mass: 114.82 u
- Electron Configuration: $[Kr]5s^2 4d^{10} 5p^1$
- Oxidation States: +3
- Electronegativity (Pauling Scale): 1.78
- Atomic Radius (van der Waals): 193 pm
- Ionization Energy: 5.786 eV
- Electron Affinity: 0.3 eV
- Melting Point: 429.75 K
- Boiling Point: 2345 K
- Density: 7.31 g/cm^3

History & Characteristics

- Year Discovered: 1863
- Discovered by: Ferdinand Reich and Hieronymous Richter
- Appearance: Silvery-white
- Natural Occurrence: Primordial
- Quick Fact: Indium is a technology-critical element, and is used to fill the growing global need for LCDs (liquid crystal displays).

Tin

Atomic Number: 50 **Atomic Symbol:** Sn
Cat: Post-Transition Metal **Stand. State:** Solid

Physical & Atomic Properties

- Atomic Mass: 118.71 u
- Electron Configuration: [Kr]$5s^2 4d^{10} 5p^2$
- Oxidation States: +4, +2
- Electronegativity (Pauling Scale): 1.96
- Atomic Radius (van der Waals): 217 pm
- Ionization Energy: 7.344 eV
- Electron Affinity: 1.2 eV
- Melting Point: 505.08 K
- Boiling Point: 2875 K
- Density: 7.287 g/cm^3

History & Characteristics

- Year Discovered: Ancient
- Discovered by: Various ancient civilizations
- Appearance: Silvery-white
- Natural Occurrence: Primordial
- Quick Fact: Tin is used to prevent corrosion and produce glass, and it's most often found alloyed with other metals. Pewter, for example, is mostly tin.

Antimony

Atomic Number: 51 **Atomic Symbol:** Sb
Category: Metalloid **Standard State:** Solid

Physical & Atomic Properties

- Atomic Mass: 121.76 u
- Electron Configuration: $[Kr]5s^2 4d^{10} 5p^3$
- Oxidation States: +5, +3, -3
- Electronegativity (Pauling Scale): 2.05
- Atomic Radius (van der Waals): 206 pm
- Ionization Energy: 8.64 eV
- Electron Affinity: 1.07 eV
- Melting Point: 903.78 K
- Boiling Point: 1860 K
- Density: 6.685 g/cm^3

History & Characteristics

- Year Discovered: Ancient
- Discovered by: Various ancient civilizations
- Appearance: Silvery-white
- Natural Occurrence: Primordial
- Quick Fact: An alloy of antimony and lead is used in batteries, bullets, cable coverings, and chemical equipment like tanks, pipes, and pumps.

Tellurium

Atomic Number: 52 **Atomic Symbol:** Te
Category: Metalloid **Standard State:** Solid

Physical & Atomic Properties

- Atomic Mass: 127.6 u
- Electron Configuration: $[Kr]5s^2 4d^{10} 5p^4$
- Oxidation States: +6, +4, -2
- Electronegativity (Pauling Scale): 2.1
- Atomic Radius (van der Waals): 206 pm
- Ionization Energy: 9.010 eV
- Electron Affinity: 1.971 eV
- Melting Point: 722.66 K
- Boiling Point: 1261 K
- Density: 6.232 g/cm^3

History & Characteristics

- Year Discovered: 1782
- Discovered by: Franz Joseph Müller von Reichenstein
- Appearance: Silvery (crystalline) or brown-black powder
- Natural Occurrence: Primordial
- Quick Fact: Tellurium is used to improve the machinability of copper and stainless steel.

Iodine

Atomic Number: 53 **Atomic Symbol:** I
Category: Halogen **Standard State:** Solid

Physical & Atomic Properties

- Atomic Mass: 126.9045 u
- Electron Configuration: [Kr]$5s^2 4d^{10} 5p^5$
- Oxidation States: +7, +5, +1, -1
- Electronegativity (Pauling Scale): 2.66
- Atomic Radius (van der Waals): 198 pm
- Ionization Energy: 10.451 eV
- Electron Affinity: 3.059 eV
- Melting Point: 386.85 K
- Boiling Point: 457.55 K
- Density: 4.93 g/cm^3

History & Characteristics

- Year Discovered: 1811
- Discovered by: Bernard Courtois
- Appearance: Bluish-black solid; purple vapor
- Natural Occurrence: Primordial
- Quick Fact: Iodine is the heaviest essential mineral nutrient and is used in the synthesis of thyroid hormones. Iodine deficiency affects about two billion people worldwide.

Xenon

Atomic Number: 54 **Atomic Symbol:** Xe
Category: Noble Gas **Standard State:** Gas

Physical & Atomic Properties

- Atomic Mass: 131.29 u
- Electron Configuration: $[Kr]5s^2 4d^{10} 5p^6$
- Oxidation States: 0
- Electronegativity (Pauling Scale): 2.6
- Atomic Radius (van der Waals): 216 pm
- Ionization Energy: 12.130 eV
- Melting Point: 161.36 K
- Boiling Point: 165.03 K
- Density: 0.005887 g/cm^3

History & Characteristics

- Year Discovered: 1898
- Discovered by: William Ramsay and Morris W. Travers
- Appearance: Colorless
- Natural Occurrence: Primordial
- Quick Fact: Xenon is used in photographic flashes, in arc lamps for motion picture projection, and in high pressure arc lamps to produce ultraviolet light.

Cesium

Atomic Number: 55 **Atomic Symbol:** Cs
Category: Alkali Metal **Standard State:** Solid

Physical & Atomic Properties

- Atomic Mass: 132.9054520 u
- Electron Configuration: [Xe]$6s^1$
- Oxidation States: +1
- Electronegativity (Pauling Scale): 0.79
- Atomic Radius (van der Waals): 343 pm
- Ionization Energy: 3.894 eV
- Electron Affinity: 0.472 eV
- Melting Point: 301.59 K
- Boiling Point: 944 K
- Density: 1.93 g/cm^3

History & Characteristics

- Year Discovered: 1860
- Discovered by: Robert Bunsen and Gustav Kirchhoff
- Appearance: Silvery-yellow
- Natural Occurrence: Primordial
- Quick Fact: Cesium is incredibly accurate at timekeeping and is used in atomic clocks. As a clock it loses 1 second per 100 million years.

Barium

Atomic Number: 56 **Atomic Symbol:** Ba
Category: Alkaline Earth Metal **Stand. State:** Solid

Physical & Atomic Properties

- Atomic Mass: 137.33 u
- Electron Configuration: [Xe]$6s^2$
- Oxidation States: +2
- Electronegativity (Pauling Scale): 0.89
- Atomic Radius (van der Waals): 268 pm
- Ionization Energy: 5.212 eV
- Melting Point: 1000 K
- Boiling Point: 2170 K
- Density: 3.62 g/cm^3

History & Characteristics

- Year Discovered: 1772
- Discovered by: Carl Wilhelm Scheele
- Appearance: Silvery-white
- Natural Occurrence: Primordial
- Quick Fact: Compounds containing barium have a wide range of uses; they're found in rat poison, weighting agents in oil drilling fluids, and the white fluid used to visualize intestines in an X-ray diagnostic test.

Lanthanum

Atomic Number: 57 **Atomic Symbol:** La
Category: Lanthanide **Standard State:** Solid

Physical & Atomic Properties

- Atomic Mass: 138.9055 u
- Electron Configuration: [Xe]$6s^2 5d^1$
- Oxidation States: +3
- Electronegativity (Pauling Scale): 1.1
- Atomic Radius (van der Waals): 240 pm
- Ionization Energy: 5.577 eV
- Electron Affinity: 0.5 eV
- Melting Point: 1191 K
- Boiling Point: 3737 K
- Density: 6.15 g/cm^3

History & Characteristics

- Year Discovered: 1839
- Discovered by: Carl Gustaf Mosander
- Appearance: Silvery-white
- Natural Occurrence: Primordial
- Quick Fact: Lanthanum compounds are used as catalysts, additives in glass, carbon arc lamps for studio lights and projectors, and ignition elements in lighters and torches.

Cerium

Atomic Number: 58 **Atomic Symbol:** Ce
Category: Lanthanide **Standard State:** Solid

Physical & Atomic Properties

- Atomic Mass: 140.12 u
- Electron Configuration: [Xe]$6s^2 4f^1 5d^1$
- Oxidation States: +4, +3
- Electronegativity (Pauling Scale): 1.12
- Atomic Radius (van der Waals): 235 pm
- Ionization Energy: 5.539 eV
- Electron Affinity: 0.5 eV
- Melting Point: 1071 K
- Boiling Point: 3697 K
- Density: 6.770 g/cm^3

History & Characteristics

- Year Discovered: 1803
- Discovered by: Martin Heinrich Klaproth, Jöns Jacob Berzelius, Wilhelm Hisinger
- Appearance: Gray
- Natural Occurrence: Primordial
- Quick Fact: Cerium oxide is used in incandescent gas mantles, as a glass polishing agent, and as a catalyst in self-cleaning ovens.

Praseodymium

Atomic Number: 59 **Atomic Symbol:** Pr
Category: Lanthanide **Standard State:** Solid

Physical & Atomic Properties

- Atomic Mass: 140.9077 u
- Electron Configuration: [Xe]$6s^2 4f^3$
- Oxidation States: +3
- Electronegativity (Pauling Scale): 1.13
- Atomic Radius (van der Waals): 239 pm
- Ionization Energy: 5.464 eV
- Melting Point: 1204 K
- Boiling Point: 3793 K
- Density: 6.77 g/cm^3

History & Characteristics

- Year Discovered: 1885
- Discovered by: Carl Auer von Welsbach
- Appearance: Silvery-white
- Natural Occurrence: Primordial
- Quick Fact: Praseodymium is used as an alloying agent with magnesium to create high-strength metals used in aircraft engines.

Neodymium

Atomic Number: 60 **Atomic Symbol:** Nd
Category: Lanthanide **Standard State:** Solid

Physical & Atomic Properties

- Atomic Mass: 144.24 u
- Electron Configuration: [Xe]$6s^2 4f^4$
- Oxidation States: +3
- Electronegativity (Pauling Scale): 1.14
- Atomic Radius (van der Waals): 229 pm
- Ionization Energy: 5.525 eV
- Melting Point: 1294 K
- Boiling Point: 3347 K
- Density: 7.01 g/cm^3

History & Characteristics

- Year Discovered: 1885
- Discovered by: Carl Auer von Welsbach
- Appearance: Silvery-white
- Natural Occurrence: Primordial
- Quick Fact: Neodymium compounds were first commercially used as glass dyes in 1927, and they remain a popular additive in glasses.

Promethium

Atomic Number: 61 **Atomic Symbol:** Pm
Category: Lanthanide **Standard State:** Solid

Physical & Atomic Properties

- Atomic Mass: 144.91276 u
- Electron Configuration: [Xe]$6s^2 4f^5$
- Oxidation States: +3
- Atomic Radius (van der Waals): 236 pm
- Ionization Energy: 5.55 eV
- Melting Point: 1315 K
- Boiling Point: 3273 K
- Density: 7.26 g/cm^3

History & Characteristics

- Year Discovered: 1942
- Discovered by: Chien Shiung Wu, Emilio Segrè, and Hans Albrecht Bethe
- Appearance: Silvery-white
- Natural Occurrence: From decay
- Quick Fact: Promethium is the only rare-earth radioactive metal, and is used in atomic batteries for spacecraft and guided missiles.

Samarium

Atomic Number: 62 **Atomic Symbol:** Sm
Category: Lanthanide **Standard State:** Solid

Physical & Atomic Properties

- Atomic Mass: 150.4 u
- Electron Configuration: [Xe]$6s^2 4f^6$
- Oxidation States: +3, +2
- Electronegativity (Pauling Scale): 1.17
- Atomic Radius (van der Waals): 229 pm
- Ionization Energy: 5.644 eV
- Melting Point: 1347 K
- Boiling Point: 2067 K
- Density: 7.52 g/cm^3

History & Characteristics

- Year Discovered: 1879
- Discovered by: Paul-Émile Lecoq de Boisbaudran
- Appearance: Silvery-white
- Natural Occurrence: Primordial
- Quick Fact: The leading commercial application of samarium is in samarium–cobalt magnets, which have permanent magnetization.

Europium

Atomic Number: 63 **Atomic Symbol:** Eu
Category: Lanthanide **Standard State:** Solid

Physical & Atomic Properties

- Atomic Mass: 151.96 u
- Electron Configuration: [Xe]$6s^2 4f^7$
- Oxidation States: +3, +2
- Atomic Radius (van der Waals): 233 pm
- Ionization Energy: 5.670 eV
- Melting Point: 1095 K
- Boiling Point: 1802 K
- Density: 5.24 g/cm^3

History & Characteristics

- Year Discovered: 1901
- Discovered by: Eugène-Anatole Demarçay
- Appearance: Silvery-white
- Natural Occurrence: Primordial
- Quick Fact: Europium is the most reactive of the rare-earth metals, quickly oxidizing in air and reacting quickly and vigorously with water.

Gadolinium

Atomic Number: 64 **Atomic Symbol:** Gd
Category: Lanthanide **Standard State:** Solid

Physical & Atomic Properties

- Atomic Mass: 157.2 u
- Electron Configuration: [Xe]$6s^2 4f^7 5d^1$
- Oxidation States: +3
- Electronegativity (Pauling Scale): 1.2
- Atomic Radius (van der Waals): 237 pm
- Ionization Energy: 6.150 eV
- Melting Point: 1586 K
- Boiling Point: 3546 K
- Density: 7.90 g/cm^3

History & Characteristics

- Year Discovered: 1880
- Discovered by: Jean Charles Galissard de Marignac
- Appearance: Silvery-white
- Natural Occurrence: Primordial
- Quick Fact: Gadolinium compounds are hosts for phosphors for fluorescent lamps, X-ray intensifying screens, and scintillators for X-ray tomography, and as an MRI contrast agent.

Terbium

Atomic Number: 65 **Atomic Symbol:** Tb
Category: Lanthanide **Standard State:** Solid

Physical & Atomic Properties

- Atomic Mass: 158.92535 u
- Electron Configuration: $[Xe]6s^2 4f^9$
- Oxidation States: +3
- Atomic Radius (van der Waals): 221 pm
- Ionization Energy: 5.864 eV
- Melting Point: 1629 K
- Boiling Point: 3503 K
- Density: 8.23 g/cm^3

History & Characteristics

- Year Discovered: 1843
- Discovered by: Carl Gustaf Mosander
- Appearance: Silvery-gray
- Natural Occurrence: Primordial
- Quick Fact: Terbium compounds are used as green phosphors in fluorescent lamps, computer monitors, and television screens that use cathode-ray tubes.

Dysprosium

Atomic Number: 66 **Atomic Symbol:** Dy
Category: Lanthanide **Standard State:** Solid

Physical & Atomic Properties

- Atomic Mass: 162.50 u
- Electron Configuration: [Xe]$6s^2 4f^{10}$
- Oxidation States: +3
- Electronegativity (Pauling Scale): 1.22
- Atomic Radius (van der Waals): 229 pm
- Ionization Energy: 5.939 eV
- Melting Point: 1685 K
- Boiling Point: 2840 K
- Density: 8.55 g/cm^3

History & Characteristics

- Year Discovered: 1886
- Discovered by: Paul-Émile Lecoq de Boisbaudran
- Appearance: Silvery-white
- Natural Occurrence: Primordial
- Quick Fact: Dysprosium is used for its high thermal neutron absorption cross-section in making control rods in nuclear reactors.

Holmium

Atomic Number: 67 **Atomic Symbol:** Ho
Category: Lanthanide **Standard State:** Solid

Physical & Atomic Properties

- Atomic Mass: 164.93033 u
- Electron Configuration: [Xe]$6s^2 4f^{11}$
- Oxidation States: +3
- Electronegativity (Pauling Scale): 1.23
- Atomic Radius (van der Waals): 216 pm
- Ionization Energy: 6.022 eV
- Melting Point: 1747 K
- Boiling Point: 2973 K
- Density: 8.80 g/cm^3

History & Characteristics

- Year Discovered: 1878
- Discovered by: Per Teodor Cleve, Marc Delafontaine and Jacques-Louis Soret
- Appearance: Silvery-white
- Natural Occurrence: Primordial
- Quick Fact: Holmium has the highest magnetic permeability of any element and therefore is used for the polepieces of the strongest static magnets.

Erbium

Atomic Number: 68 **Atomic Symbol:** Er
Category: Lanthanide **Standard State:** Solid

Physical & Atomic Properties

- Atomic Mass: 167.26 u
- Electron Configuration: [Xe]$6s^2 4f^{12}$
- Oxidation States: +3
- Electronegativity (Pauling Scale): 1.24
- Atomic Radius (van der Waals): 235 pm
- Ionization Energy: 6.108 eV
- Melting Point: 1802 K
- Boiling Point: 3141 K
- Density: 9.07 g/cm^3

History & Characteristics

- Year Discovered: 1843
- Discovered by: Carl Gustaf Mosander
- Appearance: Silvery-white
- Natural Occurrence: Primordial
- Quick Fact: Erbium is used in various nuclear and metallurgical applications. For example, when added to vanadium, erbium lowers the hardness and improves pliability of the metal.

Thulium

Atomic Number: 69 **Atomic Symbol:** Tm
Category: Lanthanide **Standard State:** Solid

Physical & Atomic Properties

- Atomic Mass: 168.93422 u
- Electron Configuration: [Xe]$6s^2 4f^{13}$
- Oxidation States: +3
- Electronegativity (Pauling Scale): 1.25
- Atomic Radius (van der Waals): 227 pm
- Ionization Energy: 6.184 eV
- Melting Point: 1818 K
- Boiling Point: 2223 K
- Density: 9.32 g/cm^3

History & Characteristics

- Year Discovered: 1879
- Discovered by: Per Teodor Cleve
- Appearance: Silvery-gray
- Natural Occurrence: Primordial
- Quick Fact: Despite its high price and rarity, thulium is used as the radiation source in portable X-ray devices, and in some solid-state lasers.

Ytterbium

Atomic Number: 70 **Atomic Symbol:** Yb
Category: Lanthanide **Standard State:** Solid

Physical & Atomic Properties

- Atomic Mass: 173.04 u
- Electron Configuration: [Xe]$6s^2 4f^{14}$
- Oxidation States: +3, +2
- Atomic Radius (van der Waals): 242 pm
- Ionization Energy: 6.254 eV
- Melting Point: 1092 K
- Boiling Point: 1469 K
- Density: 6.90 g/cm^3

History & Characteristics

- Year Discovered: 1878
- Discovered by: Jean Charles Galissard de Marignac
- Appearance: Silvery-white
- Natural Occurrence: Primordial
- Quick Fact: Ytterbium is added to stainless steel to improve its mechanical properties, and ytterbium fiber laser amplifiers are used in marking and engraving.

Lutetium

Atomic Number: 71 **Atomic Symbol:** Lu
Category: Lanthanide **Standard State:** Solid

Physical & Atomic Properties

- Atomic Mass: 174.967 u
- Electron Configuration: $[Xe]6s^2 4f^{14} 5d^1$
- Oxidation States: +3
- Electronegativity (Pauling Scale): 1.27
- Atomic Radius (van der Waals): 221 pm
- Ionization Energy: 5.426 eV
- Melting Point: 1936 K
- Boiling Point: 3675 K
- Density: 9.84 g/cm^3

History & Characteristics

- Year Discovered: 1907
- Discovered by: Carl Auer von Welsbach, Charles James, and Georges Urbain
- Appearance: Silvery-white
- Natural Occurrence: Primordial
- Quick Fact: Stable lutetium can be used as catalysts in petroleum cracking in refineries and can be used in alkylation, hydrogenation, and polymerization applications.

Hafnium

Atomic Number: 72 **Atomic Symbol:** Hf
Category: Transition Metal **Stand. State:** Solid

Physical & Atomic Properties

- Atomic Mass: 178.5 u
- Electron Configuration: $[Xe]6s^2 4f^{14} 5d^2$
- Oxidation States: +4
- Electronegativity (Pauling Scale): 1.3
- Atomic Radius (van der Waals): 212 pm
- Ionization Energy: 6.825 eV
- Melting Point: 2506 K
- Boiling Point: 4876 K
- Density: 13.3 g/cm^3

History & Characteristics

- Year Discovered: 1923
- Discovered by: Dirk Coster and George Charles de Hevesy
- Appearance: Silvery
- Natural Occurrence: Primordial
- Quick Fact: Hafnium is used in filaments and electrodes, and some semiconductor fabrication processes use its oxide for integrated circuits at 45 nm and smaller feature lengths.

Tantalum

Atomic Number: 73
Atomic Symbol: Ta
Category: Transition Metal
Stand. State: Solid

Physical & Atomic Properties

- Atomic Mass: 180.9479 u
- Electron Configuration: [Xe]$6s^2 4f^{14} 5d^3$
- Oxidation States: +5
- Electronegativity (Pauling Scale): 1.5
- Atomic Radius (van der Waals): 217 pm
- Ionization Energy: 7.89 eV
- Electron Affinity: 0.322 eV
- Melting Point: 3290 K
- Boiling Point: 5731 K
- Density: 16.4 g/cm^3

History & Characteristics

- Year Discovered: 1802
- Discovered by: Anders Gustaf Ekenberg
- Appearance: Gray
- Natural Occurrence: Primordial
- Quick Fact: The chemical inertness of tantalum makes it a valuable substance for laboratory equipment and a substitute for platinum.

Tungsten

Atomic Number: 74
Atomic Symbol: W
Category: Transition Metal
Stand. State: Solid

Physical & Atomic Properties

- Atomic Mass: 183.8 u
- Electron Configuration: [Xe]$6s^2 4f^{14} 5d^4$
- Oxidation States: +6
- Electronegativity (Pauling Scale): 2.36
- Atomic Radius (van der Waals): 210 pm
- Ionization Energy: 7.98 eV
- Electron Affinity: 0.815 eV
- Melting Point: 3695 K
- Boiling Point: 5828 K
- Density: 19.3 g/cm^3

History & Characteristics

- Year Discovered: 1783
- Discovered by: Juan José Elhuyar and Fausto Elhuyar
- Appearance: Silvery-white
- Natural Occurrence: Primordial
- Quick Fact: Tungsten has the highest melting point, boiling point, and tensile strength of any element.

Rhenium

Atomic Number: 75 **Atomic Symbol:** Re
Category: Transition Metal **Stand. State:** Solid

Physical & Atomic Properties

- Atomic Mass: 186.21 u
- Electron Configuration: [Xe]$6s^2 4f^{14} 5d^5$
- Oxidation States: +7, +6, +4
- Electronegativity (Pauling Scale): 1.9
- Atomic Radius (van der Waals): 217 pm
- Ionization Energy: 7.88 eV
- Electron Affinity: 0.15 eV
- Melting Point: 3459 K
- Boiling Point: 5869 K
- Density: 20.8 g/cm^3

History & Characteristics

- Year Discovered: 1925
- Discovered by: Ida Tacke-Noddack, Walter Noddack and Otto Berg
- Appearance: Silvery-white
- Natural Occurrence: Primordial
- Quick Fact: Nickel-based alloys of rhenium are used in the combustion chambers, turbine blades, and exhaust nozzles of jet engines.

Osmium

Atomic Number: 76 **Atomic Symbol:** Os
Category: Transition Metal **Stand. State:** Solid

Physical & Atomic Properties

- Atomic Mass: 190.2 u
- Electron Configuration: $[Xe]6s^2 4f^{14} 5d^6$
- Oxidation States: +4, +3
- Electronegativity (Pauling Scale): 2.2
- Atomic Radius (van der Waals): 216 pm
- Ionization Energy: 8.7 eV
- Electron Affinity: 1.1 eV
- Melting Point: 3306 K
- Boiling Point: 5285 K
- Density: 22.59 g/cm^3

History & Characteristics

- Year Discovered: 1803
- Discovered by: Smithson Tennant
- Appearance: Bluish-white
- Natural Occurrence: Primordial
- Quick Fact: Osmium is the densest naturally occurring element, with an experimentally measured (using X-ray crystallography) density of 22.59 g/cm^3.

Iridium

Atomic Number: 77 **Atomic Symbol:** Ir
Category: Transition Metal **Stand. State:** Solid

Physical & Atomic Properties

- Atomic Mass: 192.22 u
- Electron Configuration: $[Xe]6s^2 4f^{14} 5d^7$
- Oxidation States: +4, +3
- Electronegativity (Pauling Scale): 2.2
- Atomic Radius (van der Waals): 202 pm
- Ionization Energy: 9.1 eV
- Electron Affinity: 1.565 eV
- Melting Point: 2719 K
- Boiling Point: 4701 K
- Density: 22.42 g/cm^3

History & Characteristics

- Year Discovered: 1803
- Discovered by: Smithson Tennant
- Appearance: Silvery-white
- Natural Occurrence: Primordial
- Quick Fact: Iridium is the most corrosion-resistant metal, even at temperatures as high as 2000 °C.

Platinum

Atomic Number: 78 **Atomic Symbol:** Pt
Category: Transition Metal **Stand. State:** Solid

Physical & Atomic Properties

- Atomic Mass: 195.08 u
- Electron Configuration: $[Xe]6s^1 4f^{14} 5d^9$
- Oxidation States: +4, +2
- Electronegativity (Pauling Scale): 2.28
- Atomic Radius (van der Waals): 209 pm
- Ionization Energy: 9 eV
- Electron Affinity: 2.128 eV
- Melting Point: 2041.55 K
- Boiling Point: 4098 K
- Density: 21.46 g/cm^3

History & Characteristics

- Year Discovered: 1735
- Discovered by: Antonio de Ulloa
- Appearance: Silvery-white
- Natural Occurrence: Primordial
- Quick Fact: Because of its scarcity in Earth's crust, only a few hundred tonnes of platinum are produced annually, making it highly valuable and a major precious metal commodity.

Gold

Atomic Number: 79 **Atomic Symbol:** Au
Category: Transition Metal **Stand. State:** Solid

Physical & Atomic Properties

- Atomic Mass: 196.96657 u
- Electron Configuration: $[Xe]6s^1 4f^{14} 5d^{10}$
- Oxidation States: +3, +1
- Electronegativity (Pauling Scale): 2.54
- Atomic Radius (van der Waals): 166 pm
- Ionization Energy: 9.226 eV
- Electron Affinity: 2.309 eV
- Melting Point: 1337.33 K
- Boiling Point: 3129 K
- Density: 19.282 g/cm^3

History & Characteristics

- Year Discovered: Ancient
- Discovered by: Various ancient civilizations
- Appearance: Golden yellow
- Natural Occurrence: Primordial
- Quick Fact: A relatively rare-element, gold is a precious metal that has been used for coinage, jewelry, and other arts throughout recorded history.

Mercury

Atomic Number: 80 **Atomic Symbol:** Hg
Category: Transition Metal **Stand. State:** Liquid

Physical & Atomic Properties

- Atomic Mass: 200.59 u
- Electron Configuration: [Xe]$6s^2 4f^{14} 5d^{10}$
- Oxidation States: +2, +1
- Electronegativity (Pauling Scale): 2
- Atomic Radius (van der Waals): 209 pm
- Ionization Energy: 10.438 eV
- Melting Point: 234.32 K
- Boiling Point: 629.88 K
- Density: 13.5336 g/cm^3

History & Characteristics

- Year Discovered: Ancient
- Discovered by: Chinese, Egyptians, and Hindus (presumed)
- Appearance: Silvery-white
- Natural Occurrence: Primordial
- Quick Fact: Mercury is the only metallic element that is liquid at standard conditions for temperature and pressure, and has been used in thermometers and barometers.

Thallium

Atomic Number: 81 **Atomic Symbol:** Tl

Cat: Post-Transition Metal **Stand. State:** Solid

Physical & Atomic Properties

- Atomic Mass: 204.383 u
- Electron Configuration: [Xe]$6s^2 4f^{14} 5d^{10} 6p^1$
- Oxidation States: +3, +1
- Electronegativity (Pauling Scale): 1.62
- Atomic Radius (van der Waals): 196 pm
- Ionization Energy: 6.108 eV
- Electron Affinity: 0.2 eV
- Melting Point: 577 K
- Boiling Point: 1746 K
- Density: 11.8 g/cm^3

History & Characteristics

- Year Discovered: 1861
- Discovered by: William Crookes
- Appearance: Silvery-gray
- Natural Occurrence: Primordial
- Quick Fact: Nearly 70% of thallium production is used in the electronics industry, and the remainder is used in the pharmaceutical industry and in glass manufacturing.

Lead

Atomic Number: 82 **Atomic Symbol:** Pb
Cat: Post-Transition Metal **Stand. State:** Solid

Physical & Atomic Properties

- Atomic Mass: 207 u
- Electron Configuration: [Xe]$6s^2 4f^{14} 5d^{10} 6p^2$
- Oxidation States: +4, +2
- Electronegativity (Pauling Scale): 2.33
- Atomic Radius (van der Waals): 202 pm
- Ionization Energy: 7.417 eV
- Electron Affinity: 0.36 eV
- Melting Point: 600.61 K
- Boiling Point: 2022 K
- Density: 11.342 g/cm^3

History & Characteristics

- Year Discovered: Ancient
- Discovered by: Romans (presumed)
- Appearance: Bluish-gray
- Natural Occurrence: Primordial
- Quick Fact: Due to lead's relative abundance and low cost, it's commonly used in construction, plumbing, batteries, bullets, weights, solders, pewters, and fusible alloys.

Bismuth

Atomic Number: 83 **Atomic Symbol:** Bi
Cat: Post-Transition Metal **Stand. State:** Solid

Physical & Atomic Properties

- Atomic Mass: 208.9804 u
- Electron Configuration: $[Xe]6s^24f^{14}5d^{10}6p^3$
- Oxidation States: +5, +3
- Electronegativity (Pauling Scale): 2.02
- Atomic Radius (van der Waals): 207 pm
- Ionization Energy: 7.289 eV
- Electron Affinity: 0.946 eV
- Melting Point: 544.55 K
- Boiling Point: 1837 K
- Density: 9.807 g/cm^3

History & Characteristics

- Year Discovered: 1753
- Discovered by: Claude François Geoffroy
- Appearance: Silvery-white
- Natural Occurrence: Primordial
- Quick Fact: Bismuth is the most naturally diamagnetic element, and has one of the lowest values of thermal conductivity among metals.

Polonium

Atomic Number: 84 **Atomic Symbol:** Po
Category: Metalloid **Standard State:** Solid

Physical & Atomic Properties

- Atomic Mass: 208.98243 u
- Electron Configuration: $[Xe]6s^2 4f^{14} 5d^{10} 6p^4$
- Oxidation States: +4, +2
- Electronegativity (Pauling Scale): 2
- Atomic Radius (van der Waals): 197 pm
- Ionization Energy: 8.417 eV
- Electron Affinity: 1.9 eV
- Melting Point: 527 K
- Boiling Point: 1235 K
- Density: 9.32 g/cm^3

History & Characteristics

- Year Discovered: 1898
- Discovered by: Marie and Pierre Curie
- Appearance: Silvery-gray
- Natural Occurrence: From decay
- Quick Fact: Due to its intense radioactivity, polonium is extremely dangerous to humans and has mostly been investigated on the trace scale only.

Astatine

Atomic Number: 85 **Atomic Symbol:** At
Category: Halogen **Standard State:** Solid

Physical & Atomic Properties

- Atomic Mass: 209.98715 u
- Electron Configuration: [Xe]$6s^2 4f^{14} 5d^{10} 6p^5$
- Oxidation States: 7, 5, 3, 1, -1
- Electronegativity (Pauling Scale): 2.2
- Atomic Radius (van der Waals): 202 pm
- Ionization Energy: 9.5 eV
- Electron Affinity: 2.8 eV
- Melting Point: 575 K
- Density: 7 g/cm^3

History & Characteristics

- Year Discovered: 1940
- Discovered by: Dale R. Corson, Kenneth R. Mackenzie, and Emilio Segré
- Appearance: Presumed very dark
- Natural Occurrence: From decay
- Quick Fact: It is the rarest naturally occurring element in the Earth's crust. A pure sample has never been assembled because it would be vaporized by the heat of its own radioactivity.

Radon

Atomic Number: 86 **Atomic Symbol:** Rn
Category: Noble Gas **Standard State:** Gas

Physical & Atomic Properties

- Atomic Mass: 222.01758 u
- Electron Configuration: [Xe]$6s^2 4f^{14} 5d^{10} 6p^6$
- Oxidation States: 0
- Atomic Radius (van der Waals): 220 pm
- Ionization Energy: 10.745 eV
- Melting Point: 202 K
- Boiling Point: 211.45 K
- Density: 0.00973 g/cm^3

History & Characteristics

- Year Discovered: 1900
- Discovered by: Friedrich Ernst Dorn
- Appearance: Colorless
- Natural Occurrence: From decay
- Quick Fact: Radon gas is considered a health hazard and can accumulate in buildings and in low areas such as basements and crawl spaces. Radon gas can also dissolve and accumulate in ground water, such as wells.

Francium

Atomic Number: 87 **Atomic Symbol:** Fr
Category: Alkali Metal **Standard State:** Solid

Physical & Atomic Properties

- Atomic Mass: 223.01973 u
- Electron Configuration: [Rn]$7s^1$
- Oxidation States: +1
- Electronegativity (Pauling Scale): 0.7
- Atomic Radius (van der Waals): 348 pm
- Ionization Energy: 3.9 eV
- Electron Affinity: 0.47 eV
- Melting Point: 300 K

History & Characteristics

- Year Discovered: 1939
- Discovered by: Marguerite Perey
- Appearance: Presumed silvery-gray
- Natural Occurrence: From decay
- Quick Fact: Francium is extremely radioactive and its most stable isotope has a half-life of only 22 minutes. Because of its short half-life, bulk francium has never been viewed.

Radium

Atomic Number: 88 **Atomic Symbol:** Ra
Category: Alkaline Earth Metal **Stand. State:** Solid

Physical & Atomic Properties

- Atomic Mass: 226.02541 u
- Electron Configuration: [Rn]$7s^2$
- Oxidation States: +2
- Electronegativity (Pauling Scale): 0.9
- Atomic Radius (van der Waals): 283 pm
- Ionization Energy: 5.279 eV
- Melting Point: 973 K
- Boiling Point: 1413 K
- Density: 5 g/cm^3

History & Characteristics

- Year Discovered: 1898
- Discovered by: Marie and Pierre Curie
- Appearance: Silvery-white
- Natural Occurrence: From decay
- Quick Fact: Other than its use in nuclear medicine, radium has no commercial applications due to its radioactivity.

Actinium

Atomic Number: 89
Atomic Symbol: Ac
Category: Actinide
Standard State: Solid

Physical & Atomic Properties

- Atomic Mass: 227.02775 u
- Electron Configuration: [Rn]$7s^2 6d^1$
- Oxidation States: +3
- Electronegativity (Pauling Scale): 1.1
- Atomic Radius (van der Waals): 260 pm
- Ionization Energy: 5.17 eV
- Melting Point: 1324 K
- Boiling Point: 3471 K
- Density: 10.07 g/cm^3

History & Characteristics

- Year Discovered: 1899
- Discovered by: André-Louis Debierne
- Appearance: Silvery-white
- Natural Occurrence: From decay
- Quick Fact: Due to its scarcity, high price, and radioactivity, actinium has no significant industrial uses. Its current applications include a neutron source and an agent for radiation therapy.

Thorium

Atomic Number: 90 **Atomic Symbol:** Th
Category: Actinide **Standard State:** Solid

Physical & Atomic Properties

- Atomic Mass: 232.038 u
- Electron Configuration: [Rn]$7s^2 6d^2$
- Oxidation States: +4
- Electronegativity (Pauling Scale): 1.3
- Atomic Radius (van der Waals): 237 pm
- Ionization Energy: 6.08 eV
- Melting Point: 2023 K
- Boiling Point: 5061 K
- Density: 11.72 g/cm^3

History & Characteristics

- Year Discovered: 1828
- Discovered by: Jöns Jacob Berzelius
- Appearance: Silvery
- Natural Occurrence: Primordial
- Quick Fact: Because of its physical and atomic properties, thorium has been suggested as a replacement for uranium as nuclear fuel in nuclear reactors, and several thorium reactors have been built.

Protactinium

Atomic Number: 91 **Atomic Symbol:** Pa
Category: Actinide **Standard State:** Solid

Physical & Atomic Properties

- Atomic Mass: 231.0359 u
- Electron Configuration: $[Rn]7s^2 5f^2 6d^1$
- Oxidation States: +5, +4
- Electronegativity (Pauling Scale): 1.5
- Atomic Radius (van der Waals): 243 pm
- Ionization Energy: 5.89 eV
- Melting Point: 1845 K
- Boiling Point: 4273 K
- Density: 15.37 g/cm^3

History & Characteristics

- Year Discovered: 1913
- Discovered by: Kasimir Fajans and Oswald Helmuth Göhring
- Appearance: Silvery
- Natural Occurrence: From decay
- Quick Fact: Concentrations of protactinium in the Earth's crust are a few parts per trillion, and because of its scarcity and radioactivity, it has no uses outside scientific research.

Uranium

Atomic Number: 92 **Atomic Symbol:** U
Category: Actinide **Standard State:** Solid

Physical & Atomic Properties

- Atomic Mass: 238.0289 u
- Electron Configuration: [Rn]$7s^2 5f^3 6d^1$
- Oxidation States: +6, +5, +4, +3
- Electronegativity (Pauling Scale): 1.38
- Atomic Radius (van der Waals): 240 pm
- Ionization Energy: 6.194 eV
- Melting Point: 1408 K
- Boiling Point: 4404 K
- Density: 18.95 g/cm^3

History & Characteristics

- Year Discovered: 1789
- Discovered by: Martin Heinrich Klaproth
- Appearance: Silvery-white
- Natural Occurrence: Primordial
- Quick Fact: Many uses of uranium exploit its unique nuclear properties. Uranium-235 is the only naturally occurring fissile isotope, which makes it widely used in nuclear power plants and nuclear weapons.

Neptunium

Atomic Number: 93 **Atomic Symbol:** Np
Category: Actinide **Standard State:** Solid

Physical & Atomic Properties

- Atomic Mass: 237.04817 u
- Electron Configuration: $[Rn]7s^2 5f^4 6d^1$
- Oxidation States: +6, +5, +4, +3
- Electronegativity (Pauling Scale): 1.36
- Atomic Radius (van der Waals): 221 pm
- Ionization Energy: 6.266 eV
- Melting Point: 917 K
- Boiling Point: 4175 K
- Density: 20.25 g/cm^3

History & Characteristics

- Year Discovered: 1940
- Discovered by: Edwin McMillan and Philip H. Ableson
- Appearance: Silvery
- Natural Occurrence: From decay
- Quick Fact: Neptunium is used as a precursor for the formation of plutonium-238, used in radioisotope thermal generators to provide electricity for spacecraft.

Plutonium

Atomic Number: 94 **Atomic Symbol:** Pu
Category: Actinide **Standard State:** Solid

Physical & Atomic Properties

- Atomic Mass: 244.06420 u
- Electron Configuration: $[Rn]7s^2 5f^6$
- Oxidation States: +6, +5, +4, +3
- Electronegativity (Pauling Scale): 1.28
- Atomic Radius (van der Waals): 243 pm
- Ionization Energy: 6.06 eV
- Melting Point: 913 K
- Boiling Point: 3501 K
- Density: 19.84 g/cm^3

History & Characteristics

- Year Discovered: 1940
- Discovered by: Glenn T. Seaborg, Arthur Wahl, Joseph Kennedy, and Edwin McMillan
- Appearance: Silvery-white
- Natural Occurrence: From decay
- Quick Fact: Producing plutonium in useful quantities for the first time was a major part of the Manhattan Project during World War II that developed the first atomic bombs.

Americium

Atomic Number: 95
Atomic Symbol: Am
Category: Actinide
Standard State: Solid

Physical & Atomic Properties

- Atomic Mass: 243.06138 u
- Electron Configuration: [Rn]$7s^2 5f^7$
- Oxidation States: +6, +5, +4, +3
- Electronegativity (Pauling Scale): 1.3
- Atomic Radius (van der Waals): 244 pm
- Ionization Energy: 5.993 eV
- Melting Point: 1449 K
- Boiling Point: 2284 K
- Density: 13.69 g/cm^3

History & Characteristics

- Year Discovered: 1944
- Discovered by: Glenn T. Seaborg, Ralph A. James, Leon O. Morgan, and Albert Ghiorso
- Appearance: Silvery-white
- Natural Occurrence: Synthetic
- Quick Fact: Americium is primarily produced in nuclear reactors through the bombardment of plutonium with neutrons.

Curium

Atomic Number: 96 **Atomic Symbol:** Cm
Category: Actinide **Standard State:** Solid

Physical & Atomic Properties

- Atomic Mass: 247.07035 u
- Electron Configuration: $[Rn]7s^2 5f^7 6d^1$
- Oxidation States: +3
- Electronegativity (Pauling Scale): 1.3
- Atomic Radius (van der Waals): 245 pm
- Ionization Energy: 6.02 eV
- Melting Point: 1618 K
- Boiling Point: 3400 K
- Density: 13.51 g/cm^3

History & Characteristics

- Year Discovered: 1944
- Discovered by: Glenn T. Seaborg, Ralph A. James, and Albert Ghiorso
- Appearance: Silvery-white
- Natural Occurrence: Synthetic
- Quick Fact: Curium does not occur in nature and must be made in a nuclear reactor by neutron capture reactions from plutonium and americium isotopes.

Berkelium

Atomic Number: 97 **Atomic Symbol:** Bk
Category: Actinide **Standard State:** Solid

Physical & Atomic Properties

- Atomic Mass: 247.07031 u
- Electron Configuration: $[Rn]7s^25f^9$
- Oxidation States: +4, +3
- Electronegativity (Pauling Scale): 1.3
- Atomic Radius (van der Waals): 244 pm
- Ionization Energy: 6.23 eV
- Melting Point: 1323 K
- Density: 14 g/cm^3

History & Characteristics

- Year Discovered: 1949
- Discovered by: Stanley G. Thompson, Albert Ghiorso, and Glenn T. Seaborg
- Appearance: Silvery-white
- Natural Occurrence: Synthetic
- Quick Fact: Just over one gram of berkelium has been produced in the United States since 1967, and there is no practical application of berkelium outside scientific research.

Californium

Atomic Number: 98 **Atomic Symbol:** Cf
Category: Actinide **Standard State:** Solid

Physical & Atomic Properties

- Atomic Mass: 251.07959 u
- Electron Configuration: [Rn]$7s^2 5f^{10}$
- Oxidation States: +3
- Electronegativity (Pauling Scale): 1.3
- Atomic Radius (van der Waals): 245 pm
- Ionization Energy: 6.30 eV
- Melting Point: 1173 K

History & Characteristics

- Year Discovered: 1950
- Discovered by: Stanley G. Thompson, Glenn T. Seaborg, Kenneth Street, Jr. and Albert Ghiorso
- Appearance: Silvery-white
- Natural Occurrence: Synthetic
- Quick Fact: Californium is one of the few transuranium elements that have practical applications. Most of these applications exploit the property of certain isotopes of californium to emit neutrons.

Einsteinium

Atomic Number: 99 **Atomic Symbol:** Es
Category: Actinide **Standard State:** Solid

Physical & Atomic Properties

- Atomic Mass: 252.0830 u
- Electron Configuration: [Rn]$7s^2 5f^{11}$
- Oxidation States: +3
- Electronegativity (Pauling Scale): 1.3
- Atomic Radius (van der Waals): 245 pm
- Ionization Energy: 6.42 eV
- Melting Point: 1133 K

History & Characteristics

- Year Discovered: 1952
- Discovered by: Scientists led by Albert Ghiorso
- Appearance: Silvery
- Natural Occurrence: Synthetic
- Quick Fact: Einsteinium was discovered as a component of the debris of the first hydrogen bomb explosion in 1952, and was named after Albert Einstein.

Fermium

Atomic Number: 100

Atomic Symbol: Fm

Category: Actinide

Standard State: Solid

Physical & Atomic Properties

- Atomic Mass: 257.09511 u
- Electron Configuration: [Rn]$5f^{12}7s^2$
- Oxidation States: +3
- Electronegativity (Pauling Scale): 1.3
- Ionization Energy: 6.50 eV
- Melting Point: 1800 K

History & Characteristics

- Year Discovered: 1952
- Discovered by: Scientists led by Albert Ghiorso
- Appearance: Presumed silvery
- Natural Occurrence: Synthetic
- Quick Fact: Fermium is the heaviest element that can be formed by neutron bombardment of lighter elements, and the last element that can be prepared in macroscopic quantities, although pure fermium metal has not yet been prepared.

Mendelevium

Atomic Number: 101
Atomic Symbol: Md
Category: Actinide
Standard State: Solid

Physical & Atomic Properties

- Atomic Mass: 258.09843 u
- Electron Configuration: $[Rn]7s^2 5f^{13}$
- Oxidation States: +3, +2
- Electronegativity (Pauling Scale): 1.3
- Ionization Energy: 6.58 eV
- Melting Point: 1100 K

History & Characteristics

- Year Discovered: 1955
- Discovered by: Albert Ghiorso, Gregory R. Choppin, Bernard G. Harvey, Glenn T. Seaborg, and Stanley G. Thompson
- Appearance: Presumed silvery
- Natural Occurrence: Synthetic
- Quick Fact: Mendelevium was discovered by bombarding einsteinium with alpha particles in 1955, the same method still used to produce it today. It was named after Dmitri Mendeleev, father of the periodic table of the chemical elements.

Nobelium

Atomic Number: 102 **Atomic Symbol:** No
Category: Actinide **Standard State:** Solid

Physical & Atomic Properties

- Atomic Mass: 259.10100 u
- Electron Configuration: $[Rn]7s^2 5f^{14}$
- Oxidation States: +3, +2
- Electronegativity (Pauling Scale): 1.3
- Ionization Energy: 6.65 eV
- Melting Point: 1100 K

History & Characteristics

- Year Discovered: 1958
- Discovered by: Albert Ghiorso, Glenn T. Seaborg, Torbjørn Sikkeland, and John R. Walton
- Appearance: Presumed silvery
- Natural Occurrence: Synthetic
- Quick Fact: Nobelium was named in honor of famed chemist and philanthropist Alfred Nobel.

Lawrencium

Atomic Number: 103 **Atomic Symbol:** Lr
Category: Actinide **Standard State:** Solid

Physical & Atomic Properties

- Atomic Mass: 262.110 u
- Electron Configuration: $[Rn]7s^2 5f^{14} 6d^1$
- Oxidation States: +3
- Electronegativity (Pauling Scale): 1.3
- Melting Point: 1900 K

History & Characteristics

- Year Discovered: 1961
- Discovered by: Albert Ghiorso, Torbjørn Sikkeland, Almon E. Larsh, and Robert M. Latimer
- Appearance: Presumed silvery
- Natural Occurrence: Synthetic
- Quick Fact: Lawrencium is produced by bombarding californium placed in a linear accelerator with boron ions. It has only been produced in small quantities.

Rutherfordium

Atomic Number: 104 **Atomic Symbol:** Rf
Category: Transition Metal **Stand. State:** Solid

Physical & Atomic Properties

- Atomic Mass: 267.122 u
- Electron Configuration: [Rn]$7s^2 5f^{14} 6d^2$
- Oxidation States: +4
- Atomic Radius (van der Waals): 150 pm

History & Characteristics

- Year Discovered: 1964
- Discovered by: First discovered by Georgy Flerov and colleagues
- Appearance: Presumed silvery
- Natural Occurrence: Synthetic
- Quick Fact: Rutherfordium is a synthetic element, and because its 12 isotopes all have short half-lives, little is known about its properties. It is presumed to be a solid metal.

Dubnium

Atomic Number: 105 **Atomic Symbol:** Db
Category: Transition Metal **Stand. State:** Solid

Physical & Atomic Properties

- Atomic Mass: 268.126 u
- Electron Configuration: $[Rn]7s^25f^{14}6d^3$
- Oxidation States: 5, 4, 3
- Atomic Radius (van der Waals): 139 pm

History & Characteristics

- Year Discovered: 1967
- Discovered by: First discovered by Georgy Flerov and colleagues
- Appearance: Presumed silvery
- Natural Occurrence: Synthetic
- Quick Fact: Dubnium is named for Dubna, Russia, home of the Joint Institute for Nuclear Research, where the element was first reported.

Seaborgium

Atomic Number: 106 **Atomic Symbol:** Sg
Category: Transition Metal **Stand. State:** Solid

Physical & Atomic Properties

- Atomic Mass: 271.134 u
- Electron Configuration: [Rn]$7s^2 5f^{14} 6d^4$
- Oxidation States: 6, 5, 4, 3, 0
- Atomic Radius (van der Waals): 132 pm

History & Characteristics

- Year Discovered: 1974
- Discovered by: Scientists led by Albert Ghiorso and at the Joint Institute for Nuclear Research
- Appearance: Presumed silvery
- Natural Occurrence: Synthetic
- Quick Fact: Seaborgium was named after American nuclear chemist Glenn T. Seaborg, and is one of only two elements named after a living person at the time of naming, the other being oganesson.

Bohrium

Atomic Number: 107 **Atomic Symbol:** Bh
Category: Transition Metal **Stand. State:** Solid

Physical & Atomic Properties

- Atomic Mass: 274.144 u
- Electron Configuration: $[Rn]7s^2 5f^{14} 6d^5$
- Oxidation States: 7, 5, 4, 3
- Atomic Radius (van der Waals): 128 pm

History & Characteristics

- Year Discovered: 1976
- Discovered by: Team of scientists in Dubna, Russia
- Appearance: Presumed silvery
- Natural Occurrence: Synthetic
- Quick Fact: Bohrium was named after Danish physicist Niels Bohr, and for a time it was known as nielsbohrium with the symbol Ns.

Hassium

Atomic Number: 108 **Atomic Symbol:** Hs
Category: Transition Metal **Stand. State:** Solid

Physical & Atomic Properties

- Atomic Mass: 277.152 u
- Electron Configuration: [Rn]$7s^2 5f^{14} 6d^6$
- Oxidation States: 8, 6, 5, 4, 3, 2
- Atomic Radius (van der Waals): 126 pm

History & Characteristics

- Year Discovered: 1984
- Discovered by: Peter Armbruster and Gottfried Münzenber
- Appearance: Presumed silvery
- Natural Occurrence: Synthetic
- Quick Fact: Hassium is highly radioactive and is made by bombarding atoms of a lead isotope with ions of an iron isotope.

Meitnerium

Atomic Number: 109 **Atomic Symbol:** Mt
Category: Transition Metal **Stand. State:** Solid

Physical & Atomic Properties

- Atomic Mass: 278.156 u
- Electron Configuration: $[Rn]7s^2 5f^{14} 6d^7$
- Oxidation States: 9, 8, 6, 4, 3, 1
- Atomic Radius (van der Waals): 128 pm

History & Characteristics

- Year Discovered: 1982
- Discovered by: Scientists led by Peter Armbruster and Gottfried Münzenber
- Appearance: Presumed silvery
- Natural Occurrence: Synthetic
- Quick Fact: Meitnerium can be made by bombarding atoms of bismuth-209 with ions of iron-58 using a linear accelerator. It is calculated to have similar properties to its lighter homologues, cobalt, rhodium, and iridium.

Darmstadtium

Atomic Number: 110 **Atomic Symbol:** Ds
Category: Transition Metal **State:** Solid (Expected)

Physical & Atomic Properties

- Atomic Mass: 281.165 u
- Electron Configuration: [Rn]$7s^2 5f^{14} 6d^8$
- Oxidation States: 8, 6, 4, 2, 0
- Atomic Radius (van der Waals): 132 pm

History & Characteristics

- Year Discovered: 1994
- Discovered by: Scientists led by Peter Armbruster and Gottfried Münzenberg
- Appearance: Presumed silvery
- Natural Occurrence: Synthetic
- Quick Fact: Darmstadtium is classified as a metal and is expected to be a solid at room temperature. Its first produced atoms had a lifetime of only 1/100th of a second, but scientists have since produced heavier, more stable versions.

Roentgenium

Atomic Number: 111
Atomic Symbol: Rg
Category: Transition Metal
State: Solid (Expected)

Physical & Atomic Properties

- Atomic Mass: 282.169 u
- Electron Configuration: $[Rn]7s^2 5f^{14} 6d^9$
- Oxidation States: 5, 3, 1, -1
- Atomic Radius (van der Waals): 138 pm

History & Characteristics

- Year Discovered: 1994
- Discovered by: Scientists led by Sigurd Hofmann
- Appearance: Presumed silvery
- Natural Occurrence: Synthetic
- Quick Fact: Roentgenium is produced artificially by bombarding atoms of bismuth with ions of nickel in a linear accelerator. Only a few atoms of roentgenium have ever been made, and they have no current application outside of scientific study.

Copernicium

Atomic Number: 112 **Atomic Symbol:** Cn
Category: Transition Metal **State:** Solid (Expected)

Physical & Atomic Properties

- Atomic Mass: 285.177 u
- Electron Configuration: [Rn]$7s^2 5f^{14} 6d^{10}$
- Oxidation States: 2, 1, 0
- Atomic Radius (van der Waals): 147 pm

History & Characteristics

- Year Discovered: 1996
- Discovered by: Scientists led by Sigurd Hofmann
- Appearance: Presumed silvery
- Natural Occurrence: Synthetic
- Quick Fact: Copernicium is produced artificially by bombarding atoms of lead with ions of zinc through a linear accelerator. It was named after the famed astronomer Nicolaus Copernicus.

Nihonium

Atomic Number: 113 **Atomic Symbol:** Nb
Cat: Post-Transition Metal **State:** Solid (Expected)

Physical & Atomic Properties

- Atomic Mass: 286.183 u
- Electron Configuration: [Rn]$5f^{14}6d^{10}7s^{2}7p^{1}$
- Atomic Radius (van der Waals): 170 pm

History & Characteristics

- Year Discovered: 2004 (first undisputed claim)
- Discovered by: Kosuke Morita and colleagues
- Appearance: Presumed silvery
- Natural Occurrence: Synthetic
- Quick Fact: Nihonium has been calculated to have similar properties to its homologues boron, aluminum, gallium, indium, and thallium, but very little is known about nihonium because it's only been made in very small amounts that decay away within seconds.

Flerovium

Atomic Number: 114 **Atomic Symbol:** Fl
Cat: Post-Transition Metal **State:** Solid (Expected)

Physical & Atomic Properties

- Atomic Mass: 289.191 u
- Electron Configuration: $[Rn]7s^2 7p^2 5f^{14} 6d^{10}$
- Oxidation States: 6, 4, 2, 1, 0
- Atomic Radius (van der Waals): 180 pm

History & Characteristics

- Year Discovered: 1998
- Discovered by: Scientists led by Yuri Oganessian
- Appearance: Presumed silvery
- Natural Occurrence: Synthetic
- Quick Fact: About 90 atoms of flerovium have been observed: 58 were synthesized directly, and the rest were made from the radioactive decay of heavier elements. All of these flerovium atoms have been shown to have mass numbers from 284 to 290.

Moscovium

Atomic Number: 115 **Atomic Symbol:** Mc

Cat: Post-Transition Metal **State:** Solid (Expected)

Physical & Atomic Properties

- Atomic Mass: 290.196 u
- Electron Configuration: [Rn]$7s^2 7p^3 5f^{14} 6d^{10}$
- Oxidation States: 3, 1
- Atomic Radius (van der Waals): 187 pm

History & Characteristics

- Year Discovered: 2003
- Discovered by: Scientists led by Yuri Oganessian
- Appearance: Presumed silvery
- Natural Occurrence: Synthetic
- Quick Fact: About 100 atoms of moscovium have been observed to date, all of which have been shown to have mass numbers from 287 to 290.

Livermorium

Atomic Number: 116 **Atomic Symbol:** Lv
Cat: Post-Transition Metal **State:** Solid (Expected)

Physical & Atomic Properties

- Atomic Mass: 293.205 u
- Electron Configuration: $[Rn]7s^{2}7p^{4}5f^{14}6d^{10}$
- Oxidation States: +4, +2, -2
- Atomic Radius (van der Waals): 183 pm

History & Characteristics

- Year Discovered: 2000
- Discovered by: Scientists at the Joint Institute for Nuclear Research
- Appearance: Presumed silvery
- Natural Occurrence: Synthetic
- Quick Fact: Four isotopes of livermorium are known, with mass numbers between 290 and 293; the longest-lived among them is livermorium-293 with a half-life of about 60 milliseconds.

Tennessine

Atomic Number: 117
Atomic Symbol: Ts
Category: Halogen
State: Solid (Expected)

Physical & Atomic Properties

- Atomic Mass: 294.211 u
- Electron Configuration: $[Rn]7s^27p^55f^{14}6d^{10}$
- Oxidation States: +5, +3, +1, -1
- Atomic Radius (van der Waals): 138 pm

History & Characteristics

- Year Discovered: 2010
- Discovered by: Scientists at the Joint Institute for Nuclear Research
- Appearance: Presumed dark
- Natural Occurrence: Synthetic
- Quick Fact: Tennessine may be located in the "island of stability," a concept that explains why some superheavy elements are more stable compared to an overall trend of decreasing stability for elements beyond bismuth on the periodic table. The synthesized tennessine atoms have lasted tens and hundreds of milliseconds.

Oganesson

Atomic Number: 118 **Atomic Symbol:** Og
Category: Noble Gas **State:** Gas (Expected)

Physical & Atomic Properties

- Atomic Mass: 294.214 u
- Electron Configuration: [Rn]$7s^2 7p^6 5f^{14} 6d^{10}$
- Oxidation States: +6, +4, +2, +1, 0, -1

History & Characteristics

- Year Discovered: 2006
- Discovered by: Scientists from the Joint Institute for Nuclear Research
- Appearance: Presumed colorless
- Natural Occurrence: Synthetic
- Quick Fact: Oganesson has the highest atomic number and highest atomic mass of all known elements. It was named after nuclear physicist Yuri Oganessian, and is one of only two elements named after a living person at the time of naming, the other being seaborgium.

Index

Germanium (Ge): 32
Gold (Au): 79

H

Hafnium (Hf): 72
Hassium (Hs): 108
Helium (He): 2
Holmium (Ho): 67
Hydrogen (H): 1

I – K

Indium (In): 49
Iodine (I) 53
Iridium (Ir): 77
Iron (Fe): 26
Krypton (Kr): 36

L

Lanthanum (La): 57
Lawrencium (Lr): 103
Lead (Pb): 82
Lithium (Li): 3
Livermorium (Lv): 116
Lutetium (Lu): 71

M

Magnesium (Mg): 12
Manganese (Mn): 25
Meitnerium (Mt): 109
Mendelevium (Md): 101
Mercury (Hg): 80
Molybdenum (Mo): 42
Moscovium (Mc): 115

N

Neodymium (Nd): 60
Neon (Ne): 10
Neptunium (Np): 93
Nickel (Ni): 28
Nihonium (Nh): 113
Niobium (Nb): 41
Nitrogen (N): 7
Nobelium (No): 102

O

Oganesson (Og): 118
Osmium (Os): 76
Oxygen (O): 8

P – Q

Palladium (Pd): 46
Phosphorus (P): 15
Platinum (Pt): 78
Plutonium (Pu): 94
Polonium (Po): 84
Potassium (K): 19
Praseodymium (Pr): 59
Promethium (Pm): 61

Protactinium (Pa): 91

R

Radium (Ra): 88
Radon (Rn): 86
Rhenium (Re): 75
Rhodium (Rh): 45
Roentgenium (Rg): 111
Rubidium (Rb): 37
Ruthenium (Ru): 44
Rutherfordium (Rf): 104

S

Samarium (Sm): 62
Scandium (Sc): 21
Seaborgium (Sg): 106
Selenium (Se): 34
Silicon (Si): 14
Silver (Ag): 47
Sodium (Na): 11
Strontium (Sr): 38
Sulfur (S): 16

T

Tantalum (Ta): 73
Technetium (Tc): 43
Tellurium (Te): 52
Tennessine (Ts): 117
Terbium (Tb): 65
Thallium (Tl): 81
Thorium (Th): 90
Thulium (Tm): 69
Tin (Sn): 50
Titanium (Ti): 22
Tungsten (W): 74

U – Z

Uranium (U): 92
Vanadium (V): 23
Xenon (Xe): 54
Ytterbium (Yb): 70
Yttrium (Y): 39
Zinc (Zn): 30
Zirconium (Zr): 40

Made in United States
Cleveland, OH
22 May 2025

17118444R10069